IDENTIFICATION PROBLEMS IN THE SOCIAL SCIENCES

Identification Problems in the Social Sciences

Charles F. Manski

HARVARD UNIVERSITY PRESS
CAMBRIDGE, MASSACHUSETTS
LONDON, ENGLAND 1995

Library of Congress Cataloging-in-Publication Data

Manski, Charles F.
  Identification problems in the social sciences / Charles F.
  Manski.
    p.   cm.
  Includes bibliographical references and index.
  ISBN 0-674-44283-0
  1. Social sciences—Statistical methods.   2. Estimation theory.   I. Title.
HA29.M2465   1995
300'.5'51—dc20
94-29313
  CIP

For Ben and Becca

Contents

Preface

As a new Ph.D. student almost a quarter of a century ago, I was introduced to the study of identification by Franklin Fisher, who devoted a large part of the required course in econometrics to the problem of identification of linear simultaneous equations models. Some of my research from the early 1970s through the middle 1980s concerned the identification of discrete response models. But the perspective on identification presented in this book took shape more recently.

In the spring of 1987, Irving Piliavin stimulated me to examine the basic probabilistic structure of the selection problem associated with survey nonresponse. Later, after assuming the directorship of the Institute for Research on Poverty, I found myself grappling with the selection and mixing problems that arise in the evaluation of social programs and with the reflection problem faced in the study of neighborhood effects. My research on these subjects yielded a stream of new findings, and I gradually came to see a set of themes that warranted development into a book. Throughout this period of discovery I had the immense good fortune of being able to discuss ideas with, and receive encouragement from, my colleague Arthur Goldberger.

In 1992 I was invited by the American Sociological Association to present some of my work on identification at its annual meeting and to write a synthesis article for *Sociological Methodology* (Manski, 1993a). I am grateful to Clifford Clogg, Robert Mare, Peter Marsden, and William Mason for their roles in creating this opportunity for me to organize the disparate strands of my research into a more coherent whole.

With the article completed, writing a book seemed less daunting. The perfect environment for composing a first draft was provided by the Center for Advanced Studies in the Behavioral Sciences, where I was able to work with close to total concentration during the academic year 1992–93. I appreciate the financial support that made this possible, obtained from National Science Foundation grant SES90-22192 and from the University of Wisconsin Graduate School.

Arthur Goldberger, Jeffrey Dominitz, and Elizabeth Uhr have offered many constructive substantive comments and editorial suggestions on a draft of the manuscript. It has been a pleasure to work with Michael Aronson and Elizabeth Gretz of Harvard University Press.

June 1994

IDENTIFICATION PROBLEMS
IN THE SOCIAL SCIENCES

Introduction

Suppose that you observe the almost simultaneous movements of a man and his image in a mirror. Does the mirror image cause the man's movements or reflect them? If you do not understand something of optics and human behavior, you will not be able to tell.

A like inferential problem arises if you try to interpret the common observation that individuals belonging to the same group tend to behave similarly. Two hypotheses often advanced to explain this phenomenon are

> *endogenous effects,* wherein the propensity of an individual to behave in some way varies with the prevalence of that behavior in the group

and

> *correlated effects,* wherein individuals in the same group tend to behave similarly because they face similar institutional environments or have similar individual characteristics.

Similar behavior within groups could stem from endogenous effects; for example, group members could experience pressure to conform to group norms. Or group similarities might reflect correlated effects; for example, persons with similar characteristics might choose to associate with one another. If you do not know something about the way groups form and the way their members interact, then you cannot distinguish between these hypotheses.

Why might you care whether observed patterns of behavior are generated by endogenous effects, by correlated effects, or in some other way? A good practical reason is that different processes have differing implications for public policy. For example, understanding how students interact in classrooms is critical to the evaluation of many aspects of educational policy, from ability tracking to class-size standards to racial integration programs.

Suppose that, unable to interpret observed patterns of behavior, you seek the expert advice of two social scientists. One, perhaps a sociologist, asserts that pressure to conform to group norms makes the individuals in a group tend to behave similarly. The other, perhaps an economist, asserts that persons with similar characteristics choose to associate with one another. Both assertions are consistent with the empirical evidence, so you have no objective way to assess their validity. All you can do is judge the persuasiveness of the arguments offered. If you are persuaded by one social scientist more than by the other, it is only because one is a more skilled advocate for his or her position.

The situation just depicted is frustratingly familiar. There are many good reasons to want to know why the members of groups tend to behave similarly. Nevertheless, researchers have been unable to resolve the question.

Social scientists rarely seem able to settle questions of public concern. Consider, for example, the never-ending American debate about Aid to Families with Dependent Children (AFDC), the social insurance program commonly referred to as welfare. A central issue is the effect of AFDC on marriage, fertility, and labor supply behavior. Almost everyone has an opinion on the matter, but the opinions vary widely. Researchers have worked hard to understand how individuals respond to the incentives embedded in AFDC (see Moffitt, 1992a, and Manski and Garfinkel, 1992). But disagreements about the behavioral effects of welfare persist.

Social scientists have similarly worked hard to understand how the threat of punishment deters crime (see Blumstein, Cohen, and Nagin, 1978), how class size and composition affect student learning (see Hanushek, 1986, and Gamoran, 1992), how neighborhoods affect their inhabitants (see Jencks and Mayer, 1989), and how family structure affects children's outcomes (see Hayes and Hofferth, 1987, and

McLanahan and Sandefur, 1994). In these and so many other areas, progress is painfully slow. Research accumulates but does not converge toward a consensus.

Why do social scientists so often provide conflicting perspectives on questions of public interest? The core problem is the inherent difficulty of studying human behavior. The conclusions that can be drawn from any analysis are determined by the assumptions made and by the data brought to bear. The range of plausible assumptions about human behavior is wide. The available data are limited to observations that can be made without undue intrusion.[1] Researchers combining limited data with different maintained assumptions can, and often do, reach different logically valid conclusions.

A contributing problem is the frequent failure of social scientists to face up to the difficulty of their enterprise. Researchers sometimes do not recognize that the interpretation of data requires assumptions. Researchers sometimes understand the logic of scientific inference but ignore it when reporting their own work. The scientific community rewards those who produce strong novel findings. The public, impatient for solutions to its pressing concerns, rewards those who offer simple analyses leading to unequivocal policy recommendations. These incentives make it tempting for researchers to maintain assumptions far stronger than they can persuasively defend, in order to draw strong conclusions.

Identification

Methodological research is concerned with the logic of scientific inference. The objective is to learn what conclusions can and cannot be drawn given specified combinations of assumptions and data.

Empirical researchers usually enjoy learning of positive methodological findings. Particularly pleasing are results showing that conventional assumptions, when combined with available data, imply stronger conclusions than previously recognized. Negative findings are less welcome. Researchers are especially reluctant to learn that, given the available data, some conclusion of interest cannot be drawn unless strong assumptions are invoked. Be this as it may, both positive and negative findings are important to the advancement of science.

For over a century, methodological research in the social sciences has made productive use of statistical theory.[2] One supposes that the empirical problem is to infer some feature of a population described by a probability distribution and that the available data are observations extracted from the population by some sampling process. One combines the data with assumptions about the population and the sampling process to draw statistical conclusions about the population feature of interest.

Working within this framework, it is useful to separate the inferential problem into statistical and identification components. Studies of identification seek to characterize the conclusions that could be drawn if one could use the sampling process to obtain an unlimited number of observations. Studies of statistical inference seek to characterize the generally weaker conclusions that can be drawn from a finite number of observations.

Statistical and identification problems limit in distinct ways the conclusions that may be drawn in empirical research. Statistical problems may be severe in small samples but diminish in importance as the sampling process generates more observations. Identification problems cannot be solved by gathering more of the same kind of data. These inferential difficulties can be alleviated only by invoking stronger assumptions or by initiating new sampling processes that yield different kinds of data.

To illustrate the distinction, consider Figures I.1 and I.2. Both figures concern a researcher who wants to predict a random variable y conditional on a specified value for some other variable x. The available data are a random sample of observations of (y, x) drawn from a population in which x only takes values in the intervals $[0, 4]$ and $[6, 8]$. In Figure I.1, the researcher has 100 observations of (y, x) and uses these data to draw a confidence interval for the expected value of y conditional on x. In Figure I.2, the researcher has 1000 observations and similarly draws a confidence interval.

Inspect the intervals $[0, 4]$ and $[6, 8]$. The wide confidence interval of Figure I.1 is a statistical problem. Gathering more data permits one to estimate the conditional expectation of y more precisely and so narrows the confidence interval, as shown in Figure I.2. Now inspect the interval $(4, 6)$. The confidence interval is infinitely wide in Figure I.1 and remains so in Figure I.2. This is an identification problem. The sampling process generates no observations in the interval $(4, 6)$,

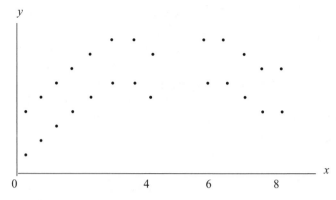

Figure I.1   Confidence interval based on 100 observations.

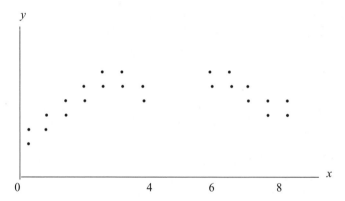

Figure I.2   Confidence interval based on 1000 observations.

so the researcher cannot possibly infer the expected value of *y* there.

It is tempting to connect the segments found in intervals [0, 4] and [6, 8] to cover (4, 6) as well. The researcher could do this if he or she were willing to assume that the expected value of *y* varies linearly with *x* over the entire interval [0, 8]. But the researcher might assume instead that the expected value of *y* remains constant as *x* varies between 4 and 6; perhaps it stays midway between its values at *x* = 4 and *x* = 6. With the available data, there is no objective way to extrapolate the segments.

Extrapolation is a particularly common and familiar identification problem. Distinguishing endogenous effects from correlated effects is another identification problem. A classic identification problem in economics is confronted when one tries to use data on market-equilibrium prices and quantities transacted to infer the supply behavior of firms and the demand behavior of consumers.

The American debate about the incentive effects of welfare also stems from an identification problem. It has been observed that the marriage, fertility, and labor supply behavior of welfare recipients tends to differ from that of nonrecipients. Unless one knows a good bit about human behavior, one cannot tell whether welfare programs influence recipients to behave in certain ways or whether individuals who behave in those ways choose to receive welfare.

These and other identification problems in the social sciences are the subject of this book. Empirical research must, of course, contend with statistical issues as well as with identification problems. Nevertheless, the two types of inferential difficulties are sufficiently distinct for it to be fruitful to study them separately. The study of identification logically comes first. Negative identification findings imply that statistical inference is fruitless: it makes no sense to try to use a sample of finite size to infer something that could not be learned even if a sample of infinite size were available. Positive identification findings imply that one should go on to study the feasibility of statistical inference.

The usefulness of separating the identification and statistical components of inference has long been recognized. Koopmans (1949, p. 132) put it this way in the article that introduced the term *identification* into the literature:

> In our discussion we have used the phrase "a parameter that can be determined from a sufficient number of observations." We shall now define this concept more sharply, and give it the ~~name~~ *identifiability* of a parameter. Instead of reasoning, as before, from "a sufficiently large number of observations" we shall base our discussion on a hypothetical knowledge of the probability distribution of the observations, as defined more fully below. It is clear that exact knowledge of this probability distribution cannot be derived from any finite number of observations. Such knowledge is the limit approachable but not attainable by extended observation. By hypothesizing nevertheless the full availabil-

ity of such knowledge, we obtain a clear separation between problems of statistical inference arising from the variability of finite samples, and problems of identification in which we explore the limits to which inference even from an infinite number of observations is suspect.

I focus on identification problems that arise when we attempt to make conditional predictions; that is, when we attempt to answer questions of the form "What if?" This book examines the conditional predictions that can and cannot be made given specified assumptions and empirical evidence.

Not all research is concerned with prediction, so focusing the book on prediction does influence its content. Scientists sometimes conduct research as an effort to improve our "understanding" of a subject, and they argue that this is a worthwhile objective even if there are no interesting implications for prediction. For example, in a text on statistical methods in epidemiology, Fleiss (1981, p. 92) states that the retrospective studies of disease that are a staple of medical research do not yield policy-relevant predictions and so are "necessarily useless from the point of view of public health." Nevertheless, the author goes on to say that "retrospective studies are eminently valid from the more general point of view of the advancement of knowledge." Justifications of this sort will not be found in the present book.

The book contains seven chapters. Chapters 1 through 4 examine observational problems that arise in all scientific work, whether in the social or the natural sciences. The central concern is to explain how the sampling process affects the predictions that can be made. Chapters 5 through 7 examine identification problems particular to the prediction of individual behavior and social interactions. A recurring interest is to compare the distinct approaches taken by different social science disciplines.

Chapters 1 and 2 cover basic material that is often referred to subsequently. Chapters 3 through 7 can be read independent of one another. Although the book examines a wide range of identification problems, it makes no pretense of being encyclopedic. Much of the book draws on my own recent research.

Tolerating Ambiguity

In addition to analyzing specific identification problems, this book develops a general theme. Social scientists and policymakers alike seem driven to draw sharp conclusions, even when these can be generated only by imposing much stronger assumptions than can be defended. We need to develop a greater tolerance for ambiguity. We must face up to the fact that we cannot answer all of the questions that we ask.

The pressure to produce answers, without qualifications, seems particularly intense in the environs of Washington, D.C. A perhaps apocryphal, but quite believable, story circulates about an economist's attempt to describe his uncertainty about a forecast to President Lyndon B. Johnson. The economist presented his forecast as a likely range of values for the quantity under discussion. Johnson is said to have replied, "Ranges are for cattle. Give me a number."

A thoughtful news magazine article written at the height of the 1992 presidential campaign compared the predictions made by participants in the campaign to those of the fictional prophet Carnac played by the popular television comedian Johnny Carson. Whitman (1992, p. 36) wrote: "An unpublished commandment of presidential campaigns can be stated simply: Thou shalt never say, 'I don't have an answer to this crisis.' Instead, both candidates and pundits feel obliged to act out their Carnac complex in election years, professing, like Johnny Carson's famous seer, to be all seeing and all knowing."

Social scientists should recognize how hard it is to provide firm answers to complex social questions. Some scientific conventions, notably the reporting of sampling confidence intervals in statistical analysis, do promote the expression of uncertainty. But other scientific practices encourage misplaced certainty.

One problem is the fixation of social scientists on point identification of parameters. Empirical studies typically seek to learn the value of some parameter characterizing the population of interest. The conventional practice is to invoke assumptions strong enough to identify the exact value of this parameter. Even if these assumptions are implausible, they are defended as necessary for inference to proceed. Yet identification is not an all-or-nothing proposition. Weaker and more plausible assumptions often suffice to bound parameters in informative ways.

A larger problem is the common view that a scientist should choose one hypothesis to maintain, even if that means discarding others that are a priori plausible and consistent with the available empirical evidence. This view was expressed in an influential methodological essay written by Milton Friedman over forty years ago. Friedman (1953) placed prediction as the central objective of science, writing, "The ultimate goal of a positive science is the development of a 'theory' or 'hypothesis' that yields valid and meaningful (i.e. not truistic) predictions about phenomena not yet observed" (p. 5). He went on to say, "The choice among alternative hypotheses equally consistent with the available evidence must to some extent be arbitrary, though there is general agreement that relevant considerations are suggested by the criteria 'simplicity' and 'fruitfulness,' themselves notions that defy completely objective specification" (p. 10).

I do not see why a scientist must choose one hypothesis to hold, especially when this requires the use of "to some extent . . . arbitrary" criteria. Indeed, using arbitrary criteria to choose a single hypothesis has an obvious drawback in predicting phenomena not yet observed: one may have made a wrong choice. Social scientists are notorious for making sharp predictions that turn out to be incorrect. The credibility of social science would be higher if we would strive to offer predictions under the range of plausible hypotheses that are consistent with the available evidence.

1

Extrapolation

1.1. Predicting Criminality

In 1982 the RAND Corporation released a study of criminal behavior as reported in 1978 by a sample of 2,200 prison and jail inmates in California, Michigan, and Texas (Chaiken and Chaiken, 1982, and Greenwood, 1982). Most respondents reported that they had committed five or fewer crimes per year in the period before their current arrest and conviction. A small group reported much higher rates of crime commission, in some cases more than one hundred per year.

The researchers found that, using limited information on a person's past convictions, drug use, and employment, they could predict well whether that person had been a high-rate offender. This finding suggested to at least part of the research team that *selective incapacitation* should be encouraged as a crime-fighting tool (Greenwood, 1982). Selective incapacitation calls for the sentencing of convicted criminals to be tied to predictions of their future criminality. Those with backgrounds that predict high rates of offenses would receive longer prison terms than those with other backgrounds.

The RAND study generated much controversy, especially when a prediction approach devised by Greenwood found its way into legislative proposals for selective incapacitation (see Blackmore and Welsh, 1983, and Blumstein et al., 1986). Some of the controversy concerned the normative acceptability of selective incapacitation, but much of it concerned the validity of extrapolations from the RAND findings.

The findings characterize the empirical association between background and reported crime commission within one cohort of inmates imprisoned in three states under the sentencing policies then in effect.

Would this association continue to hold when applied to other cohorts of inmates in other states? Would it hold when applied to convicted criminals who have not been imprisoned under existing sentencing policies? Would it hold if sentencing policy were to change? In particular, would it hold if selective incapacitation were to be implemented?

The RAND study did not address these questions. Greenwood's approach to the prediction of criminality simply presumed that the empirical association between background and reported crime commission would remain approximately the same when extrapolated to other times, places, and sentencing policies.

1.2. Probabilistic Prediction

To move beyond a loose discussion of extrapolation, we need to be familiar with the standard probabilistic framework that social scientists, decision theorists, and statisticians use to describe prediction problems. This framework will be employed throughout the book.

Conditional Distributions

Given a specified value for some variables x, the problem is to predict the realization of a random variable y. In the RAND study of criminal behavior, for example, x included a person's past convictions, drug use, and employment. The variable y was the number of crimes of a particular type that a person commits in a specified time period. Criminologists treat the rate of crime commission as a random variable to express the idea that criminal behavior may be heterogeneous among persons with the same observed attributes. Treating crime commission as random does not require one to imagine that an actual randomizing device, such as a lottery, determines criminal behavior.

Given a value for x, the most that one can do to predict y is to determine the probability distribution of y conditional on x. This distribution will be denoted $P(y \mid x)$. Empirical research in the social sciences is fundamentally concerned with the problem of learning such conditional probability distributions.

The term "probability distribution of y conditional on x" is cumbersome, so I often refer to $P(y \mid x)$ simply as the distribution of y.[1]

I could also shorten the notation by denoting this distribution as $P(y)$, leaving the conditioning on x implicit. I do not take this step because I want the reader to keep in mind that specification of the conditioning variables is as essential to the definition of a prediction problem as is specification of the outcomes to be predicted. The inferential problems examined in this book can be addressed for any specification of y and x, but are well defined only after these variables are specified.

Best Predictors

One needs to know the entire distribution $P(y \mid x)$ to characterize the outcomes y fully, but one can make some useful predictions with more limited information. Although y is random, one might be asked to select a single number as a "best" prediction of y conditional on x. Let this number, which may depend on x, be denoted $p(x)$. Suppose that, having chosen $p(x)$, one suffers a loss $L[y - p(x)]$ whose magnitude varies with the prediction error $y - p(x)$; the further the prediction error is from zero, the larger the loss. Then one might choose $p(x)$ to minimize expected loss conditional on x. A solution to this minimization problem is called a *best predictor* of y conditional on x.[2]

The best predictor is determined by the loss function $L(\cdot)$ and by the probability distribution $P(y \mid x)$. Two especially prominent loss functions, both of which treat under- and over-predictions of y symmetrically, are square loss and absolute loss; that is, $L(u) = u^2$ and $L(u) = |u|$, where $u \equiv y - p(x)$ denotes the prediction error. The resulting best predictors are well known to be the mean of y conditional on x and the median of y conditional on x. Features of conditional probability distributions are generically referred to as *regressions;* the mean and median of y conditional on x are the mean and median regressions of y on x.[3]

One does not need to know the entire distribution $P(y \mid x)$ to predict y in the sense just described. One need only know its mean, median, or some other feature depending on the loss function. It is, however, essential to understand that the various best predictors are not the same. When one specifies a loss function and minimizes expected loss, one attempts to summarize the entire conditional probability distribution $P(y \mid x)$ by a single number. Different loss functions generate distinct summary statistics.

For example, if one changes from square loss to absolute loss, the best predictor changes from the mean to the median. Laypeople and researchers alike sometimes act as if the mean and the median are interchangeable statistics. This is incorrect. In fact, Chapter 2 examines an important situation in which one can bound the median of a distribution yet learn nothing at all about its mean.

1.3. Inferring Conditional Distributions from Random-Sample Data

How might one come to know the conditional distribution $P(y \mid x)$? The classical literature on statistical inference presumes that a researcher can draw observations at random from a population, each of whose members has a value of the pair (y, x). The joint probability distribution $P(y, x)$ describes the frequency with which different values of (y, x) appear in the population. The conditional distribution $P(y \mid x)$ describes the frequency with which different values of y appear within the subpopulation sharing any specified value of x.

Random sampling reveals $P(y, x)$ even if one has no prior information about this distribution.[4] An easy way to see this is to consider the problem of estimating the probability $P[(y, x) \in A]$ that (y, x) falls in some set A. (The symbol "$\in$" means "is an element of.") Given a random sample of N observations $[(y_i, x_i), i = 1, \ldots, N]$, a natural estimate of the probability is the fraction of observations falling in set A; that is,

$$(1.1) \qquad \frac{1}{N} \sum_{i=1}^{N} 1[(y_i, x_i) \in A].$$

The indicator function $1[\cdot]$ takes the value one if the bracketed logical condition holds, and zero otherwise. The strong law of large numbers implies that, as the sample size increases, the fraction of observations falling in A almost surely converges to $P[(y, x) \in A]$. So, with random sampling, one can learn $P[(y, x) \in A]$ even if one knew nothing a priori about its value. This holds for every set A, and therefore one can learn the distribution $P(y, x)$.

To make conditional predictions, we are interested not in

$P(y, x)$ but in $P(y \mid x)$ at a specified value of x, say x_0. There are three cases to consider. What distinguishes these cases is the status of the value x_0 in the population.

Suppose first that the value x_0 occurs with positive frequency; that is, $P(x = x_0) > 0$. Then no prior information is needed to identify the conditional distribution $P(y \mid x = x_0)$. To see this, consider the problem of estimating the probability $P(y \in B \mid x = x_0)$ that y falls in some set B, conditional on x taking the value x_0. Given a random sample of N observations, an obvious estimate is the sample frequency with which y falls in B, among those observations i for which $x_i = x_0$; that is,

$$(1.2) \qquad \frac{\displaystyle\sum_{i=1}^{N} 1[y_i \in B] \, 1[x_i = x_0]}{\displaystyle\sum_{i=1}^{N} 1[x_i = x_0]}.$$

The strong law of large numbers implies that, as the sample increases in size, the cell frequency (1.2) almost surely converges to $P(y \in B \mid x = x_0)$. This holds for every set B. So one can learn the conditional distribution $P(y \mid x = x_0)$.

Now suppose that $P(x = x_0) = 0$, but there is positive probability that x falls arbitrarily close to x_0. This is the situation when x has a continuous distribution with positive density at x_0; for example, when x is normally distributed. The cell-frequency estimate for $P(y \in B \mid x = x_0)$ no longer works; with probability one, no sample observations of x equal x_0. Values of x close to x_0, however, are likely to appear in the sample. This suggests that $P(y \in B \mid x = x_0)$ might be estimated by the sample frequency with which y falls in B, among those observations for which x_i is suitably near x_0; that is, by a *local frequency* of the form

$$(1.3) \qquad \frac{\displaystyle\sum_{i=1}^{N} 1[y_i \in B] \, 1[|x_i - x_0| < d_N]}{\displaystyle\sum_{i=1}^{N} 1[|x_i - x_0| < d_N]}.$$

When x is scalar, $|x_i - x_0|$ denotes the absolute value of the difference $x_i - x_0$. When x is a vector, $|x_i - x_0|$ denotes any reasonable measure of the distance between x_i and x_0; for example, the Euclidean distance between these vectors. The parameter d_N is a sample-size dependent *bandwidth* chosen by the researcher to operationalize the idea that one wishes to confine attention to those observations in which x_i is near x_0.

A basic finding of the modern statistical literature on *nonparametric regression* analysis is that the local-frequency estimate (1.3) almost surely converges to $P(y \in B \mid x = x_0)$ as the sample grows in size, provided that three conditions hold:

(a) $P(y \in B \mid x)$ varies continuously with x, for x near x_0.
(b) One tightens the bandwidth as the sample size increases.
(c) One does not tighten the bandwidth too rapidly as the sample size increases.

Only condition (a) presumes prior information, namely, that $P(y \in B \mid x)$ varies smoothly with x. A researcher can always choose bandwidths so that conditions (b) and (c) hold.[5]

Finally, suppose that there is zero probability that x falls near x_0. That is, there is some $d_0 > 0$ such that $P(|x - x_0| < d_0) = 0$. For example, this was the situation in the interval (4, 6) of Figures I.1 and I.2 in the Introduction. Then the local-frequency estimate (1.3) no longer works; the estimate ceases to exist when one attempts to reduce the bandwidth d_N below d_0. This failure is symptomatic of a fundamental problem. If there is zero probability of observing x near x_0, then the conditional distribution $P(y \mid x = x_0)$ can be varied arbitrarily without changing the distribution of (y, x). It follows that, in the absence of prior information, $P(y \mid x = x_0)$ is not identified.

A value x_0 is said to be on the *support* of the probability distribution of x if there is positive probability of observing x arbitrarily close to x_0. The discussion in this section has shown that data collection by random sampling enables one to learn the conditional distribution $P(y \mid x = x_0)$ at essentially any value x_0 on the support of $P(x)$; the only assumption that one might need to make is that $P(y \mid x)$ varies continuously with x. If x_0 is on the support, $P(y \mid x = x_0)$ is not only identified but easy to estimate using (1.2), (1.3), or related methods.[6] These findings are much of the reason why random sampling is held in such high regard.

A value x_0 is said to be off the support of $P(x)$ if there is zero probability of observing x within some neighborhood of x_0. *Extrapolation* is the problem of identifying $P(y \mid x = x_0)$ when x_0 is off the support.

1.4. Prior Distributional Information

Suppose that the only available empirical evidence is observation of realizations from the distribution $P(y, x)$. Then the only way to identify $P(y \mid x = x_0)$ at x_0 off the support of $P(x)$ is to impose assumptions enabling one to deduce $P(y \mid x = x_0)$ from $P(y, x)$. What kinds of prior information do and do not have identifying power?

An important negative fact is that the continuity assumption (condition (a) of Section 1.3) that enabled nonparametric estimation of $P(y \in B \mid x)$ on the support of $P(x)$ reveals nothing about $P(y \in B \mid x)$ off the support. To understand why, let x_1 be the point on the support that is closest to x_0. Continuity of $P(y \mid x)$ in x implies that $P(y \mid x)$ is near $P(y \mid x = x_1)$ when x is near x_1, but does not tell us how to interpret the two uses of the word "near" as magnitudes. In particular, we do not know whether the distance separating x_0 and x_1 should be interpreted as large or small.

Clearly this reasoning applies not only to continuity but to any smoothness assumption that only restricts the behavior of $P(y \mid x)$ locally around the point x_1.[7] Thus extrapolation must require prior information that restricts $P(y \mid x)$ globally. In other words, it requires information that enables one to deduce $P(y \mid x = x_0)$ from knowledge of $P(y \mid x)$ at x-values that are not necessarily "near" x_0.

Invariance Assumptions

Perhaps the most common practice is to assume that y behaves in the same way at x_0 as it does at some specified x_1 on the support of $P(x)$. That is, assume

$$(1.4) \quad P(y \mid x = x_0) = P(y \mid x = x_1).$$

This *invariance* assumption was implicitly invoked by Greenwood when he recommended that his predictor of criminality be applied to times, places, and sentencing policies other than those observed in the RAND study.

An invariance assumption is often applied when the outcomes of social experiments with randomized assignment of treatments are used to predict the outcomes of actual social programs. An experiment is said to have *external validity* if the distribution of outcomes realized by a treatment group is the same as the distribution of outcomes that would be realized in an actual program. Campbell and Stanley (1963, p. 5) state: "*External validity* asks the question of *generalizability:* To what populations, settings, treatment variables, and measurement variables can this effect be generalized?"

Assumption (1.4), under the name *temporal invariance,* is applied routinely in efforts to predict the future. Let $x \equiv (w, t)$, where t denotes a date in time and w denotes some covariates that do not vary with time. A familiar extrapolation problem is to determine a conditional distribution $P(y \mid w, t = t_0)$, at some date t_0 that is yet to occur. Clearly no data collection undertaken before t_0 can identify this distribution. It is therefore common to assume that history repeats itself, conditional on w. That is, one assumes that

$$(1.5) \quad P(y \mid w, t = t_0) = P(y \mid w, t = t_1),$$

where t_1 is a past date at which data collection did take place. Random sampling at date t_1 identifies $P(y \mid w, t = t_1)$ at w on the support of $P(w)$. So assumption (1.5) identifies $P(y \mid w, t = t_0)$ at these values of w.

Theory, Hypothesis Testing, and Extrapolation

From the perspective of prediction, the primary function of scientific theory is to justify the imposition of invariance and other distributional assumptions that enable extrapolation. A secondary function is to justify imposing assumptions that improve the sampling precision with which conditional distributions may be estimated on the support of the conditioning variables. The latter function of theory is secondary because prior information is not essential to estimate $P(y \mid x)$ on the support but is essential to extrapolation.

Scientists and philosophers sometimes say that a function of theory is to explain "why" observed outcomes are what they are—that is, to establish some sense of causation. This function of theory is connected to prediction only to the extent that causal explanations help motivate assumptions with predictive power. To put the matter starkly, consider Greenwood's approach to predicting criminality. A jurisdiction contemplating using the approach to make selective incapacitation decisions does not need to ask why the association between background and criminality found in the RAND study was what it was. It need only ask whether it is reasonable to assume that this association holds in its own circumstances.

The subsequent chapters of this book examine how particular social science theories have been used to motivate assumptions that enable extrapolation. I shall defer discussion of these theories until then, but a general comment about the empirical testing of theories is warranted here.

Theories are testable where they are least needed, and are not testable where they are most needed. Theories are least needed to determine conditional distributions $P(y \mid x)$ on the support of $P(x)$. They are most needed to determine these distributions off the support.

The restrictions on $P(y \mid x)$ implied by a theory may fail to hold either on or off the support of $P(x)$. Failures of a theory on the support are detectable; statisticians have developed methods of hypothesis testing for just this purpose. Failures off the support are inherently not detectable. Scientific controversies arise when researchers hypothesize different theories whose implied restrictions on $P(y \mid x)$ hold on the support of $P(x)$ but differ off the support. There is no objective way to distinguish among theories that "fit the data" but imply different extrapolations.

1.5. Predicting High School Graduation

Predicting children's outcomes conditional on family attributes is a long-standing substantive concern of the social sciences. A recent collaborative study illustrates the methodological concerns of this chapter. This study will be discussed further in Chapter 2.

Manski, Sandefur, McLanahan, and Powers (1992) used data from the National Longitudinal Survey of Youth (NLSY) to estimate the probability that an American child graduates from high school, conditional on race, sex, family structure during adolescence, and father's and mother's years of schooling. Thus we estimated $P(y = 1 \mid x)$ where $y = 1$ indicates high school graduation and where $x =$ (race, sex, family structure, parents' schooling). As in the prediction of criminality, our treatment of high school graduation as a random variable expresses the idea that outcomes may be heterogeneous among persons with the same observed attributes.

The NLSY was initiated in 1979 with a national sample of men and women aged 14 to 21. Respondents have been reinterviewed yearly since then. Our measure of family structure was a binary variable, taken from the 1979 survey, indicating whether the respondent resided in an intact or nonintact family at age 14. A nonintact family was defined to be one that does not include both biological or adoptive parents; that is, a family with one parent, with a parent and stepparent, or with no parents. High school graduation was a binary variable, taken from the 1985 survey, indicating whether a respondent received a high school diploma or GED certificate by age 20. The findings discussed here concern the samples of respondents who were aged 14 to 17 in 1979 and for whom complete data were available.

The conditioning variables in this study were all discrete, so cell frequencies of the form (1.2) could have been used to estimate the conditional probabilities of high school graduation. We thought it reasonable to assume, however, that among children with the same race, sex, and family structure, the probability of high school graduation should vary smoothly with parents' years of schooling. So, within each (race, sex, family structure) cell, we used local frequencies similar to (1.3) to estimate graduation probabilities as a function of parents' schooling.

Table 1.1 presents the findings for children whose parents have both completed twelve years of schooling. Similar estimates may be computed for children whose parents have other levels of schooling.[8]

Inspection of the table indicates several patterns. Conditioning on (sex, family structure, parents' schooling), the estimated graduation probabilities of black and white children are pretty much identical, except that a black female in a nonintact family is more likely to grad-

Table 1.1   Estimated high school graduation probabilities for children whose parents have completed twelve years of schooling

	Intact family	Nonintact family
White male	.89	.77
White female	.94	.79
Black male	.87	.78
Black female	.95	.88

Source: Manski et al. (1992), table 4.

uate than is a white female in a nonintact family. Conditioning on (race, family structure, parents' schooling), a female is more likely to graduate than is a male. Conditioning on (race, sex, parents' schooling), a child in an intact family is more likely to graduate than is a child in a nonintact family.

The entries in Table 1.1 and analogous estimates computed for other levels of parents' schooling enable one to predict high school graduation outcomes conditional on specified family attributes in the American environment of the 1980s. The NLSY data suffice to make these predictions. No theory of family dynamics, intergenerational mobility, gender roles, or race is needed. Nor need one know the graduation requirements, attendance policies, or educational philosophies of U.S. schools.

Suppose, however, that one wished to predict the outcomes that would occur if the American environment of the 1980s were to change in some way. For example, what would graduation probabilities be if American schools were to modify the graduation requirements, attendance policies, or educational philosophies that were in effect in the 1980s? What would they be if employers and postsecondary institutions were to become unwilling to accept the GED certificate as an educational credential? What would they be if social welfare agencies were to implement proposed "family-preservation" programs aimed at increasing the fraction of intact families in the population? These questions call for extrapolations and so cannot be answered using the NLSY data alone.

2

The Selection Problem

2.1. The Nature of the Problem

Nonresponse is a perennial concern in the collection of survey data. Some persons are not interviewed and some who are interviewed do not answer some of the questions posed. For example, the U.S. Bureau of the Census (1991, pp. 387–388) reported that in the March 1990 administration of its quarterly Current Population Survey (CPS), approximately 4.5 percent of the 60,000 households in the targeted sample were not interviewed. Incomplete income data were obtained from approximately 8 percent of the persons in the households that were interviewed.

Longitudinal surveys experience nonresponse in the form of sample attrition. Piliavin and Sosin (1988) interviewed a sample of 137 individuals who were homeless in Minneapolis in late December 1985. Six months later, the researchers attempted to reinterview these respondents but succeeded in locating only 78.

Survey nonresponse raises the problem of inference from censored data. Social scientists confront censoring in many other ways as well. We routinely ask *treatment-effect* questions of the form:

What is the effect of _____ on _____?

For example,

What is the effect of schooling on wages?
What is the effect of welfare recipiency on labor supply?
What is the effect of sentencing policy on crime commission?

All efforts to infer treatment effects must confront the fact that the data are inherently censored. One wants to compare outcomes across different treatments, but each unit of analysis, whether a survey respondent or an experimental subject, experiences only one of the treatments.

Whereas the implications of censoring were not well appreciated twenty years ago, they are much better understood today. In particular, social scientists have devoted substantial attention to the *selection problem*. This is the problem of identifying conditional probability distributions from random sample data in which the realizations of the conditioning variables are always observed but the realizations of outcomes are censored.

For example, a classic problem in labor economics is to learn how market wages vary with schooling, work experience, and demographic background. Available surveys such as the CPS provide background data for each respondent and wage data for those respondents who work. Even if all subjects respond fully to the questions posed, there remains a censoring problem in that the surveys do not provide market wage data for respondents who do not work. Labor economists confront the selection problem whenever they attempt to use the CPS or similar surveys to estimate wage regressions.

The selection problem is logically separate from the extrapolation problem discussed in Chapter 1. The extrapolation problem examined there arises from the fact that random sampling does not yield observations of y *off* the support of x. The selection problem arises when a censored random sampling process does not fully reveal the behavior of y *on* the support of x. So selection presents new challenges beyond those faced in extrapolation.

To introduce the selection problem formally, we need to expand the description of the population given in Chapter 1 to include a binary variable indicating when outcomes are observed. Let each member of the population be characterized by a triple (y, z, x). As before, y is the outcome to be predicted and the conditioning variables are denoted by x. The new variable z takes the value one if y is observed and the value zero otherwise.

As in Section 1.3, let a random sample be drawn from the population. One does not, however, observe all the realizations of (y, z, x). One observes all realizations of (z, x), but observes y only when $z = 1$.

This censored-sampling process does not identify $P(y \mid x)$ on the support of x. To isolate the difficulty, use the law of total probability to write $P(y \mid x)$ as the sum

$$(2.1) \quad P(y \mid x) = P(y \mid x, z = 1) P(z = 1 \mid x)$$
$$+ P(y \mid x, z = 0) P(z = 0 \mid x).$$

The censored-sampling process identifies the *selection probability* $P(z = 1 \mid x)$, the *censoring probability* $P(z = 0 \mid x)$, and the distribution $P(y \mid x, z = 1)$ of y conditional on selection. These distributions can be estimated in the manner discussed in Section 1.3. But the sampling process is uninformative regarding the distribution $P(y \mid x, z = 0)$ of y conditional on censoring. Hence the censored-sampling process reveals only that

$$(2.2) \quad P(y \mid x) = P(y \mid x, z = 1) P(z = 1 \mid x) + \gamma P(z = 0 \mid x),$$

for some unknown probability distribution γ.

The logical starting point for investigation of the selection problem is to characterize the problem in the absence of prior information about the distribution of (y, z, x); that is, to learn what restrictions on $P(y \mid x)$ are implied by (2.2) alone. Section 2.2 analyzes this *worst-case* scenario and Section 2.3 provides an empirical illustration. Section 2.4 describes the identifying power of various forms of prior information. Sections 2.5 through 2.8 examine in some depth the problem of identifying treatment effects.[1]

2.2. Identification from Censored Samples Alone

What can and cannot be learned about a conditional distribution from censored data alone, in the absence of prior information restricting the form of this distribution or the censoring rule? I first present two negative facts and then develop a set of positive findings.

Two Negative Facts

It is often assumed in empirical studies that the censored outcomes have the same distribution as the observed ones, conditional on x. That is,

$$(2.3) \quad P(y \mid x, z = 0) = P(y \mid x, z = 1).$$

This assumption, variously described as *exogenous* or *ignorable* selection, identifies $P(y \mid x)$ when combined with the empirical evidence.[2] In particular, it implies that $P(y \mid x)$ coincides with the observable distribution $P(y \mid x, z = 1)$.

Suppose a researcher asserts assumption (2.3). Can this assumption be refuted empirically? The answer is negative in the absence of prior information about the form of $P(y \mid x)$. Censored data reveal nothing about $P(y \mid x, z = 0)$, so assumption (2.3) is necessarily consistent with the empirical evidence.

Exogenous selection is an empirically testable hypothesis only if one maintains assumptions restricting the form of $P(y \mid x)$. In that case, setting $\gamma = P(y \mid x, z = 1)$ in equation (2.2) may yield an infeasible value for $P(y \mid x)$. If so, then one can conclude that assumption (2.3) must be incorrect.

The second negative fact is that, in the absence of prior information, censoring makes it impossible to learn anything about the expected value $E(y \mid x)$ of y conditional on x. To see this, write $E(y \mid x)$ as the sum

$$(2.4) \quad E(y \mid x) = E(y \mid x, z = 1) P(z = 1 \mid x)$$
$$+ E(y \mid x, z = 0) P(z = 0 \mid x).$$

The censored-sampling process identifies $E(y \mid x, z = 1)$, $P(z = 1 \mid x)$, and $P(z = 0 \mid x)$, but provides no information on $E(y \mid x, z = 0)$, which might take any value between minus and plus infinity. Hence, whenever the censoring probability $P(z = 0 \mid x)$ is positive, the available empirical evidence does not restrict the value of $E(y \mid x)$.

Bounds on Conditional Probabilities

These negative results do not imply that the selection problem is fatal in the absence of prior information. In fact, censored data do imply informative and easily interpretable bounds on important features of $P(y \mid x)$.

Let B denote any set of possible outcomes and consider the probability $P(y \in B \mid x)$ that y falls in the set B. Write this probability as the sum

$$(2.5) \quad P(y \in B \mid x) = P(y \in B \mid x, z = 1) P(z = 1 \mid x)$$
$$+ P(y \in B \mid x, z = 0) P(z = 0 \mid x).$$

The censored-sampling process identifies $P(y \in B \mid x, z = 1)$, $P(z = 1 \mid x)$, and $P(z = 0 \mid x)$, but provides no information on $P(y \in B \mid x, z = 0)$. The last quantity, however, necessarily lies between zero and one. This yields the following bound on $P(y \in B \mid x)$:

$$(2.6) \quad P(y \in B \mid x, z = 1) P(z = 1 \mid x)$$
$$\leq P(y \in B \mid x)$$
$$\leq P(y \in B \mid x, z = 1) \, P(z = 1 \mid x) + P(z = 0 \mid x).$$

The lower bound is the value $P(y \in B \mid x)$ takes if the censored values of y never fall in B, while the upper bound is the value $P(y \in B \mid x)$ takes if all the censored y fall in B.

The bound (2.6) is *sharp*. That is, nothing further can be learned about $P(y \in B \mid x)$ from censored data, in the absence of prior information about the distribution of (y, z, x). The width of the bound is the censoring probability $P(z = 0 \mid x)$. So the bound is informative unless y is always censored. Observe that the bound width may vary with x but not with the set B.[3]

It is often convenient to characterize the probability distribution of a real-valued outcome by its distribution function; that is, by the probability that y falls below different cutoff points. Make B the set of numbers less than or equal to a specified cutoff point t. Then the bound (2.6) becomes[4]

(2.7) $P(y \leq t \mid x, z = 1) P(z = 1 \mid x)$

$\leq P(y \leq t \mid x)$

$\leq P(y \leq t \mid x, z = 1) P(z = 1 \mid x) + P(z = 0 \mid x).$

It may seem surprising that one should be able to bound the distribution function of y but not its mean. The explanation is a fact central to the field of robust statistics: the mean of a random variable is not a continuous function of its distribution function. Hence small perturbations in a distribution function can generate large movements in the mean. See Huber (1981).[5]

Statistical Inference

The selection problem is a failure of identification. To keep attention focused on identification, I have treated as known the conditional distributions identified by the censored-sampling process. So the bounds have been expressed as functions of $P(y \mid x, z = 1)$ and $P(z \mid x)$. In practice, one would estimate the relevant features of these distributions, thereby obtaining estimates of the bounds. For example, to estimate the bound (2.6) on a conditional probability $P(y \in B \mid x)$, one could estimate $P(y \in B \mid x, z = 1)$ and $P(z = 1 \mid x)$ in the manner described in Section 1.3.

The precision of an estimate of the bound can be measured in the usual way by placing a confidence interval around the estimate. It is important to understand the distinction between the bound and a confidence interval around its estimate. The bound on $P(y \in B \mid x)$ is a population concept, expressing what could be learned about $P(y \in B \mid x)$ if one knew $P(y \in B \mid x, z = 1)$ and $P(z \mid x)$. The confidence interval is a sampling concept, expressing the precision with which the bound is estimated when estimates of $P(y \in B \mid x, z = 1)$ and $P(z \mid x)$ are obtained from a sample of fixed size. The confidence interval is typically wider than the bound but narrows to match the bound as the sample size increases.

2.3. Bounding the Probability of Exiting Homelessness

To illustrate the bound on conditional probabilities, consider the attrition problem that arose in the study of homelessness undertaken by Piliavin and Sosin (1988). These researchers wished to learn the probability that an individual who is homeless at a given date has a home six months later. Thus the population of interest is the set of people who are homeless at the initial date. The outcome variable y is binary, with $y = 1$ if the individual has a home six months later and $y = 0$ if the person remains homeless. The covariates x are individual background attributes. The objective is to learn $P(y = 1 \mid x)$. The censoring problem is that only a subset of the people in the original sample could be located six months later. So $z = 1$ if a respondent was located for reinterview, $z = 0$ otherwise.

Manski (1989) estimates the bound conditioning on various covariates. Suppose first that the only conditioning variable is a respondent's sex. Consider the males. Initial interview data were obtained from 106 men, of whom 64 were located six months later. Of the latter group, 21 had exited from homelessness. So the estimate of $P(y = 1 \mid \text{male}, z = 1)$ is 21/64 and that of $P(z = 1 \mid \text{male})$ is 64/106. Hence the estimate of the bound on $P(y = 1 \mid \text{male})$ is [21/106, 63/106] or approximately [.20, .59].

Now consider the females. Data were obtained from 31 women, of whom 14 were located six months later. Of these, 3 had exited from homelessness. So the estimate of $P(y = 1 \mid \text{female}, z = 1)$ is 3/14 and the estimate of $P(z = 1 \mid \text{female})$ is 14/31. Hence the estimate of the bound on $P(y = 1 \mid \text{female})$ is [3/31, 20/31], or approximately [.10, .65].

Interpretation of these estimates should be cautious, given the small sample sizes. Taking the results at face value, we have a tighter bound on $P(y = 1 \mid \text{male})$ than on $P(y = 1 \mid \text{female})$ because the attrition rate for men is lower than that for women. The attrition rates $P(z = 0 \mid x)$, hence estimated bound widths, are .39 for men and .55 for women. The important point is that both bounds are informative. Having imposed no restrictions on the attrition process, we are nevertheless able to place meaningful bounds on the probability that a person who is homeless on a given date is no longer homeless six months later.

The foregoing illustrates estimation of the bound when the conditioning variable is discrete. To provide an example with a continuous conditioning variable, let $x =$ (sex, income). The income variable is the respondent's response, expressed in dollars per week, to the question "What was the best job you ever had? How much did that job pay?"

Usable responses to the income question were obtained from 89 men and 22 women. The sample of women is too small to allow meaningful nonparametric regression analysis, so I shall restrict attention to the men. To keep the analysis simple, I ignore the additional censoring problem implied by the fact that 17 of the 106 men did not respond to the income question.

Figure 2.1 shows a local frequency estimate of the attrition probability $P(z = 0 \mid x)$, computed at the actual income values appearing in the sample. Observe that the estimated attrition probability increases smoothly over the income range where the data are concentrated but seems to turn downward in the high income range where the data are sparse.[6] Figure 2.2 graphs the estimated bound on $P(y = 1 \mid x)$. The lower bound is the estimate of $P(y = 1 \mid x, z = 1) P(z = 1 \mid x)$, which is flat on the income range where the data are concentrated but seems to turn downward eventually. The upper bound is the sum of the estimates for $P(y = 1 \mid x, z = 1) P(z = 1 \mid x)$ and for $P(z = 0 \mid x)$.

Observe that the estimated bound is tightest at the low end of the income domain and spreads as income increases. The interval is [.24, .55] at income \$50 and [.23, .66] at income \$600. This spreading reflects the fact, shown in Figure 2.1, that the estimated probability of attrition increases with income.

Is the Cup Part Empty or Part Full?

Consider the bound estimate for the probability that a male exits homelessness: that is, [.20, .59]. Even ignoring sampling variability in the estimate, this seems a modest finding. After all, $P(y = 1 \mid \text{male})$

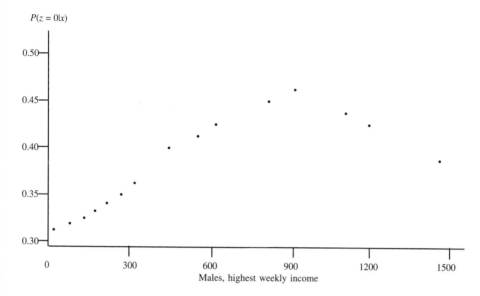

Figure 2.1   Attrition probabilities. (Source: Manski, 1989, fig. 1. Reprinted by permission of the University of Wisconsin Press.)

could take any value in an interval of width .39. Surely we would like to pin the value down more tightly than that. One might be tempted to use the midpoint of the interval [.20, .59] as a point estimate of $P(y = 1 \mid male)$ but, in the absence of prior information, there is no justification for doing so.

The bound appears more useful when one focuses on the fact that it establishes a domain of consensus about the value of $P(y = 1 \mid male)$. Researchers making different assumptions about the attrition process may logically reach different conclusions about the position of the exit probability within the interval [.20, .59]. But all researchers who accept the Piliavin and Sosin data as a censored random sample must agree that, abstracting from sampling error, the exit probability is neither less than .20 nor greater than .59. It is valuable to be able to narrow the region of potential dispute from the interval [0, 1] to the interval [.20, .59].

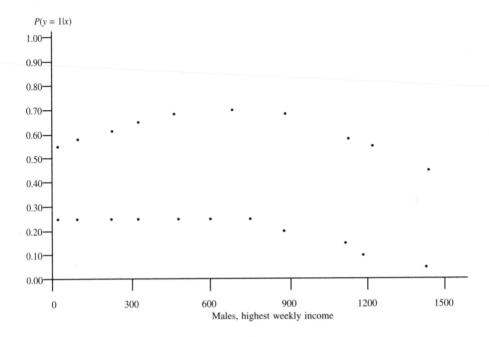

Figure 2.2   Bound on exit probabilities. (Source: Manski, 1989, fig. 2.
Reprinted by permission of the University of Wisconsin Press.)

I believe that the historical fixation of social scientists on point identification has inhibited appreciation of the usefulness of bounds. Estimable bounds on quantities that are not identified have been reported from time to time. For example, a bound developed by Frechet (1951) will be exploited in Chapter 3. But the conventional wisdom has been that bounds are uninformative.

It is relevant to note that forty years ago, in a study of the statistical problems of the Kinsey report on sexual behavior, Cochran, Mosteller, and Tukey (1954, pp. 274–282) used bounds of the form (2.6) to express the possible effects of nonresponse to the Kinsey survey. Unfortunately, the subsequent literature did not pursue the idea. In fact, Cochran (1977) dismissed bounding the effects of survey nonresponse. Using the symbol W_2 to denote the censoring probability, he stated (p. 362): "The limits are distressingly wide unless W_2 is very small." Cochran appears not to have recognized the value of worst-case bounds in establishing a domain of consensus among researchers.

2.4. Prior Distributional Information

Estimation of worst-case bounds should be the starting point for empirical analysis but ordinarily will not be the ending point. Having determined what can be learned in the absence of prior information about the distribution of (y, z, x), one should then ask what more can be learned if plausible assumptions are imposed.

Ideally, we would like to learn the identifying power of all distributional restrictions, so as to characterize the entire spectrum of inferential possibilities. But there does not appear to be any effective way to conduct an exhaustive identification analysis. I shall therefore focus on forms of prior information that are often asserted in empirical studies and that have identifying power.

It is important to understand that some kinds of prior information have no identifying power. The distribution $P(y \mid x, z = 1)$ of observed outcomes and the distribution $P(z \mid x)$ of selected and censored cases are identified by the censored-sampling process. Assumptions restricting these distributions are thus superfluous from the perspective of identification. Such assumptions may be useful from the perspective of statistical inference, as they may enable more precise estimation of $P(y \mid x, z = 1)$ and $P(z \mid x)$. But that is not our concern here.

Exogenous Selection and Self-Selection

Each year, the U.S. Bureau of the Census publishes statistics on the distribution of income in its report *Money Income of Households, Families, and Persons in the United States*. The raw data for this report are drawn from the March CPS. As noted at the beginning of the chapter, some households in the sampling frame are not interviewed and incomplete income data are obtained from some of the persons in the interviewed households. The Census Bureau copes with this nonresponse problem by assuming that, conditional on whatever variables x are observed for a given person, nonreported incomes have the same distribution as reported incomes. That is, the Census Bureau imposes the exogenous selection assumption given in equation (2.3).

Until the early 1970s, social scientists almost universally assumed exogenous selection. The perspective of many changed dramatically after economists began a sustained effort to understand the role of self-selection in determining when behavioral outcomes are observed. The result was that the assumption of exogenous selection diminished sharply in credibility.

The work of labor economists studying market wage determination has been particularly influential. Consider the process determining whether wage data are available in surveys such as the CPS. Labor economists reasoned as follows (see Gronau, 1974):

(1) Wage data are available only for respondents who work.
(2) Respondents work when they choose to do so.
(3) The wage one would be paid influences the decision to work.
(4) Hence the distributions of observed and unobserved wages may differ.

A simple model often used to express this reasoning supposes that each individual knows the wage y that would be paid if he or she were to work. The individual chooses to work if y is greater than some lowest acceptable wage R, called the person's *reservation wage,* and chooses not to work if y is below the reservation wage. So wages are observed when $y > R$ and are censored when $y < R$.

The reservation–wage model does not predict whether a person works if $y = R$, but it is conventionally assumed that this event occurs with probability zero in the population. Hence the borderline case may be ignored. With this caveat, the reservation–wage model implies that

$$(2.8) \quad P(y \mid x, z = 1) = P(y \mid x, y > R),$$

$$(2.9) \quad P(y \mid x, z = 0) = P(y \mid x, y < R),$$

and

$$(2.10) \quad P(z = 1 \mid x) = P(y > R \mid x).$$

If the reservation-wage model accurately describes the decision to work, then it is correct to assume exogenous selection if and only if

(2.11) $P(y \mid x, y > R) = P(y \mid x, y < R)$,

or equivalently, if

(2.12) $P(y \mid x, r > 0) = P(y \mid x, r < 0)$,

where $r \equiv y - R$ denotes the difference between market and reservation wages. Condition (2.12) holds if persons with high market wages also tend to have high reservation wages. In particular, it holds if the difference r between market and reservation wages is statistically independent of the market wage y. But the condition generally does not hold otherwise.

For example, suppose all persons with attributes x have the same reservation wage $R(x)$. This *fixed-threshold* assumption implies that censored wages tend to be lower than observed wages, conditional on x. In particular,

(2.13) $P[y \le t \mid x, y < R(x)] \ge P[y \le t \mid x, y > R(x)]$,

for any cutoff point t.

The reservation-wage model of labor supply provides a good example of observability determined by self-selection. There are many others that could equally well be cited. As another example, suppose one wishes to predict the outcomes a high school graduate would experience if he or she were to enroll in college (see Willis and Rosen, 1979). The outcomes of college enrollment are observable only for those high school graduates who actually enroll. The persons who enroll presumably are those who anticipate that college will have favorable outcomes for them, relative to nonenrollment. If anticipated outcomes are related to realized ones, then the observable distribution of outcomes experienced by those who actually enroll may differ from the censored distribution of outcomes that nonenrollees would have experienced had they enrolled.

The message is that self-selection may make the observability of a behavioral outcome depend partly on the value of the outcome. Self-selection does not imply that the assumption of exogenous selection is necessarily wrong, but it does cast much doubt on the assumption. Anyone concerned with prediction of y conditional on x must take notice. Using the observed distribution of outcomes to predict y conditional on x leads one astray to the extent that $P(y \mid x, z = 1)$ differs from $P(y \mid x)$.

Latent-Variable Models

It is easier to show why selection may not be exogenous than to find a credible alternative assumption that identifies $P(y \mid x)$. To illustrate, let us examine further the problem of predicting market wages. Labor economists have widely used the reservation-wage model to explain labor supply and, hence, the observability of market wages. Suppose that the reservation-wage model is correct. What is its identifying power?

The startling answer is that the reservation-wage model has no identifying power at all. The model states in equation (2.9) that the distribution of censored wages has the form $P(y \mid x, y < R)$. But the model imposes no restrictions on the form of $P(y \mid x, y < R)$. So the model implies no restrictions on the distribution of censored wages.

Facing this fact, labor economists could decide to limit themselves to estimating worst-case bounds on the distribution of market wages. Instead, the prevailing practice has been to augment the reservation-wage model by imposing assumptions on the distribution $P(y, R \mid x)$ of market and reservation wages. Because reservation wages are never observed and market wages are only sometimes observed, the reservation-wage model accompanied by restrictions on $P(y, R \mid x)$ is often referred to as a *latent-variable model*.

The identifying power of a latent-variable model depends on the assumptions imposed on $P(y, R \mid x)$. The most common practice has been to restrict $P(y, R \mid x)$ to a family of distributions specified up to some parameters. One specification, the *normal-linear model,* has received by far the most attention (see Heckman, 1976, and Maddala, 1983).

The normal-linear model assumes that, conditional on x, the distribution of $(\log y, \log R)$ is bivariate normal with mean $(x\beta_1, x\beta_2)$ and variance matrix Σ. This assumption reduces the problem of identifying $P(y \mid x)$ to that of identifying the unknown parameters $(\beta_1, \beta_2, \Sigma)$. These parameters are identified if, under the maintained log-normality assumption, there is exactly one value of $(\beta_1, \beta_2, \Sigma)$ that implies the observable distributions $P(y \mid x, z = 1)$ and $P(z \mid x)$.[7]

Some studies make the fixed-threshold assumption that all persons with attributes x have the same reservation wage $R(x)$. The magnitude of $R(x)$ need not be known a priori as it is identified by the sampling process. The reservation-wage model states that y is observed if and only if $y > R(x)$. So $R(x)$ is the lower bound of the support of the distribution $P(y \mid x, z = 1)$ of observed market wages.

The fixed-threshold assumption implies that $P(y \leq t \mid x)$ is identified at all cutoff points t greater than $R(x)$. To see this, observe that when $t > R(x)$,

$$(2.14) \quad P[y \leq t \mid x, z = 0] = P[y \leq t \mid x, y < R(x)] = 1.$$

It follows that

$$(2.15) \quad \begin{aligned} P(y \leq t \mid x) &= P(y \leq t \mid x, z = 1) P(z = 1 \mid x) \\ &\quad + P(y \leq t \mid x, z = 0) P(z = 0 \mid x) \\ &= P(y \leq t \mid x, z = 1) P(z = 1 \mid x) \\ &\quad + P(z = 0 \mid x). \end{aligned}$$

The quantities on the right side of (2.15) are all identified.

Latent-variable models with the same formal structure as the reservation-wage model have been applied to infer conditional distributions in many settings where self-selection is thought to determine the observability of outcomes. For example, economists studying the income returns to college enrollment often assume that a person enrolls in college if the income y the person would receive following enrollment exceeds the opportunity cost R of enrollment.

In the mid-1970s, the development of parametric latent-variable models of self-selection was greeted with widespread enthusiasm among economists. The initial enthusiasm was partly based on a mis-conception that specification of a latent-variable model offers a uni-versally applicable "solution" to the selection problem. Initially, it was not appreciated that solutions to the selection problem are only as good as the assumptions imposed.

Enthusiasm waned in the early 1980s after methodological studies showed that seemingly minor changes in the distributional assump-tions placed on reservation wages and similar latent variables could generate large changes in the implied value of $P(y \mid x)$. See Hurd (1979), Arabmazar and Schmidt (1982), and Goldberger (1983). Em-pirical researchers typically were unable to provide solid arguments substantiating their distributional assumptions. Hence they were forced to face up to the fragility of their inferences.

Once aware of the limitations of latent-variable models, econo-mists reacted in at least three distinct ways. Some have continued to use the models, but with a more skeptical attitude than previously. Some have returned to the earlier convention of assuming exogenous selection. Some, asserting that no useful inference is possible in the presence of censoring, have argued that controlled experimentation provides the only trustworthy basis for empirical research. The experi-mental movement will be discussed in Chapter 3.

Exclusion Restrictions

Whether they have assumed exogenous selection or used latent-variable models, researchers have generally sought to impose prior information that fully identifies the conditional distribution $P(y \mid x)$. There is a vast, mostly unexplored middle ground between the worst-case analysis of Section 2.2 and the strong assumptions consid-ered earlier in this section. This middle ground contains assumptions having some identifying power, but not enough to identify $P(y \mid x)$ fully.

One part of the middle ground that is now understood concerns the identifying power of *exclusion restrictions*. Social scientists often as-sume that some component of the regressor vector x does not affect the distribution of y, but does affect whether y is observed. For exam-ple, a labor economist studying wage determination might assume that a person's nonlabor income (for example, stock dividends and interest on savings) influences the decision to work but does not affect the wage a person would be offered by a firm.

This idea may be formalized by decomposing the conditioning variables x into two components w and v; so $x = (w, v)$. The assumption is that, holding w fixed, the outcome distribution $P(y \mid w, v)$ does not vary with v, while the selection probability $P(z = 1 \mid w, v)$ does vary with v. Thus v is excluded from the determination of y, conditional on w. The excluded component v is sometimes referred to as an *instrumental variable*.

It turns out to be easy to characterize the identifying power of an exclusion restriction. The simple result (see Manski, 1990a, 1994a) is that an exclusion restriction allows one to replace the bounds available in the absence of prior information with the intersection of these bounds across all values of v.

For example, consider the probability $P(y \in B \mid w, v)$ that y falls in some set B, conditional on (w, v). The worst-case bound on $P(y \in B \mid w, v)$ was given in equation (2.6). Now assume that, holding w fixed, $P(y \in B \mid w, v)$ does not vary with v; so $P(y \in B \mid w, v) = P(y \in B \mid w)$ for all values of v. Then $P(y \in B \mid w)$ must lie within the intersection of the bounds (2.6) on $P(y \in B \mid w, v)$ across all values of v. Thus if v can take the two values v_1 and v_2, we have

$$(2.16) \quad \max_{j=1,\,2} P[y \in B \mid (w, v_j), z = 1] P(z = 1 \mid w, v_j)$$

$$\leq P(y \in B \mid w)$$

$$\leq \min_{j=1,\,2} P[y \in B \mid (w, v_j), z = 1] P(z = 1 \mid w, v_j)$$

$$+ P(z = 0 \mid w, v_j).$$

The new bound (2.16) typically improves on (2.6), but not enough to identify $P(y \in B \mid x)$.[8]

2.5. Identification of Treatment Effects

I noted earlier that we routinely ask questions of the form: What is the effect of _____ on _____? These questions aim to compare the distributions of outcomes that would be realized if alternative treatments were applied to a population. A central problem of empirical research is to learn these distributions when some members of the population are observed to receive one treatment and the rest are observed to receive another. The remainder of the chapter examines this inferential problem.

Suppose, for example, that patients ill with a specified disease might be treated by drugs or by surgery. The relevant outcome might be life span. Then we may wish to determine the distribution of life spans that would be realized if patients with specified risk factors were all treated by drugs. And we may wish to compare this with the distribution of life spans that would be realized if these same patients were instead treated by surgery. The problem is to infer these outcome distributions from observations of the life spans of patients some of whom actually were treated by drugs and the rest by surgery.

Or, in the realm of economic policy, suppose that workers displaced by a plant closing might be retrained or given assistance in searching for a new job. The outcome of interest might be earned income. Then we may wish to determine the distribution of income that would be realized if all workers with specified backgrounds were retrained, and compare this with the distribution that would be realized if these same workers were instead assisted in job search. The problem is to infer these distributions from observations of the incomes earned by workers some of whom were retrained and some of whom were given job search assistance.

Switching Processes

To formalize the problem, let the treatments being compared be labeled 1 and 0, and let the associated outcomes be y_1 and y_0. Thus y_1 is the outcome that a person would realize if he or she were to receive treatment 1, and y_0 is the outcome that would be realized if the person were to receive treatment 0. Let $P(y_1 \mid x)$ denote the distribution of outcomes that would be realized if all persons with covariates x were to receive treatment 1, and let $P(y_0 \mid x)$ denote the analogous distribution of outcomes under treatment 0. Then the objective is to compare the distributions $P(y_1 \mid x)$ and $P(y_0 \mid x)$.[9]

We suppose that each member of the population actually receives one or the other of the two treatments. Let the binary variable z indicate which treatment a person receives. A random sample is drawn and, for each sampled person, one observes the covariate value x, the treatment z received, and the outcome under that treatment. One thus observes y_1 when $z = 1$ and y_0 when $z = 0$.

This *switching process* identifies the treatment distribution $P(z \mid x)$, the distribution $P(y_1 \mid x, z = 1)$ of y_1 conditional on receiving treatment 1, and the distribution $P(y_0 \mid x, z = 0)$ of y_0 conditional on receiving treatment 0. Thus the inferential question formalizes as:

> What does knowledge of $P(z \mid x)$, $P(y_1 \mid x, z = 1)$ and $P(y_0 \mid x, z = 0)$ reveal about $P(y_1 \mid x)$ and $P(y_0 \mid x)$?

The situations of $P(y_1 \mid x)$ and $P(y_0 \mid x)$ are symmetric, so it is enough to consider the problem of inference on $P(y_1 \mid x)$. A switching process yields richer data than did the censored-sampling process examined in Sections 2.1 through 2.4. Whereas we earlier assumed that no outcome is observed when $z = 0$, a switching process yields an observation of y_0. Whereas censored sampling identifies $P(z \mid x)$ and $P(y_1 \mid x, z = 1)$, a switching process also identifies $P(y_0 \mid x, z = 0)$. Thus a switching process differs from censored sampling to the extent that knowledge of $P(y_0 \mid x, z = 0)$ reveals something about the censored distribution $P(y_1 \mid x, z = 0)$.

Section 2.6 examines several situations in which knowing $P(y_0 \mid x, z = 0)$ does reveal something about $P(y_1 \mid x, z = 0)$. Before that, however, there are important things to say about the comparison of treatments in situations where switching and censoring are equivalent.

Comparing Treatments with No Prior Information: The Cup Is Half Full

Suppose one has no prior information about the distribution of (y_1, y_0, z, x). Then knowing $P(y_0 \mid x, z = 0)$ reveals nothing about

$P(y_1 \mid x, z = 0)$. So a switching process reveals no more about $P(y_1 \mid x)$ than does censored sampling, and the findings of Section 2.2 continue to hold. The same conclusion applies to inference on $P(y_0 \mid x)$, except that selection and censoring are reversed.

Suppose a researcher asserts that treatment has no effect on outcomes. That is, the researcher asserts that

$$(2.17) \quad P(y_1 \mid x) = P(y_0 \mid x).$$

This hypothesis is not refutable in the absence of prior information. To see this, use the law of total probability to write

$$(2.18a) \quad P(y_1 \mid x) = P(y_1 \mid x, z = 1) P(z = 1 \mid x)$$
$$+ P(y_1 \mid x, z = 0) P(z = 0 \mid x)$$

and

$$(2.18b) \quad P(y_0 \mid x) = P(y_0 \mid x, z = 1) P(z = 1 \mid x)$$
$$+ P(y_0 \mid x, z = 0) P(z = 0 \mid x).$$

A switching process reveals nothing about $P(y_1 \mid x, z = 0)$ and $P(y_0 \mid x, z = 1)$, so these distributions could equal $P(y_0 \mid x, z = 0)$ and $P(y_1 \mid x, z = 1)$ respectively. In that event, $P(y_1 \mid x)$ and $P(y_0 \mid x)$ are the same.

Although the hypothesis (2.17) is not testable, useful inferences can still be made about the relationship between $P(y_1 \mid x)$ and $P(y_0 \mid x)$. It is common to compare treatments by the difference in the probability that the outcome falls in a specified set, namely,

$$(2.19) \quad T(B \mid x) \equiv P(y_1 \in B \mid x) - P(y_0 \in B \mid x),$$

where B is the specified set of outcome values. When no prior information is available, sharp bounds on $T(B \mid x)$ can be obtained directly from the bounds on $P(y_1 \in B \mid x)$ and $P(y_0 \in B \mid x)$ given in equation (2.6). The lower bound in $T(B \mid x)$ is the difference between the lower bound on $P(y_1 \in B \mid x)$ and the upper bound on $P(y_0 = 1 \mid x)$. The upper bound on $T(B \mid x)$ is determined similarly. Hence we find that

$$(2.20) \quad P(y_1 \in B \mid x, z = 1)\,P(z = 1 \mid x)$$

$$- P(y_0 \in B \mid x, z = 0)\,P(z = 0 \mid x) - P(z = 1 \mid x)$$

$$\leq T(B \mid x)$$

$$\leq P(y_1 \in B \mid x, z = 1)\,P(z = 1 \mid x) + P(z = 0 \mid x)$$

$$- P(y_0 \in B \mid x, z = 0)\,P(z = 0 \mid x).$$

Observe that the bound (2.20) always has width one. This is an important fact with a simple explanation. The width of the bound on $T(B \mid x)$ is the sum of the widths of the bounds on $P(y_1 \in B \mid x)$ and $P(y_0 \in B \mid x)$. The latter widths are $P(z = 0 \mid x)$ and $P(z = 1 \mid x)$, respectively, as the censoring of y_1 coincides with the selection of y_0.

If no data were available, $T(B \mid x)$ could lie anywhere in the interval $[-1, 1]$, an interval of width two. So observing the outcomes of a switching process allows one to confine $T(B \mid x)$ to half of its logically possible range. In this sense, a researcher wishing to compare treatments in the absence of prior information finds that the cup is exactly half full.

Treatment Independent of Outcomes

It is often assumed that the treatment z received by each person with covariates x is statistically independent of the person's outcomes (y_1, y_0). The purpose of experimentation with randomized selection of treatment is to justify this assumption. Treatment independent of outcomes is also commonly assumed in empirical studies using nonexperimental data, where it may be referred to as *exogenous switching*

(Maddala, 1983) or as *strongly ignorable* treatment assignment (Rosenbaum and Rubin, 1983). The assumption implies that

(2.21a) $P(y_1 \mid x) = P(y_1 \mid x, z = 1) = P(y_1 \mid x, z = 0)$

and

(2.21b) $P(y_0 \mid x) = P(y_0 \mid x, z = 1) = P(y_0 \mid x, z = 0)$.

So $P(y_1 \mid x)$ and $P(y_0 \mid x)$ are identified.

When treatment is independent of outcomes, the outcome distributions $P(y_1 \mid x)$ and $P(y_0 \mid x)$ can be expressed in a manner that makes no reference to switching. Let

(2.22) $y \equiv y_1 z + y_0(1 - z)$

denote the outcome actually realized by a member of the population, namely, y_1 when $z = 1$ and y_0 otherwise. Observe that

(2.23a) $P(y \mid x, z = 1) = P(y_1 \mid x, z = 1)$

and

(2.23b) $P(y \mid x, z = 0) = P(y_0 \mid x, z = 0)$.

When treatment is independent of outcomes, (2.21) and (2.23) combine to yield

(2.24a) $P(y_1 \mid x) = P(y \mid x, z = 1)$

and

(2.24b) $P(y_0 \mid x) = P(y \mid x, z = 0)$.

Thus inference on $P(y_1 \mid x)$ and $P(y_0 \mid x)$ when the data are generated by a switching process is the same as inference on $P(y \mid x, z = 1)$ and $P(y \mid x, z = 0)$ under random sampling of (y, x, z).

Equation (2.24) can be used to rewrite the probability difference $T(B \mid x)$ of (2.19) as

$$(2.25) \quad T(B \mid x) = P(y \in B \mid x, z = 1) - P(y \in B \mid x, z = 0).$$

Empirical studies often refer to estimates of $P(y \in B \mid x, z = 1) - P(y \in B \mid x, z = 0)$ as estimates of treatment effects. This practice is well founded if treatment is independent of outcomes, but not otherwise. When treatment is not independent of outcomes, equation (2.25) generally does not hold.[10]

2.6. Information Linking Outcomes across Treatments

This section examines three distributional assumptions implying that observation of y_0 is informative about y_1, and vice versa. The situations of y_1 and y_0 are symmetric, so I focus on the problem of inferring $P(y_1 \mid x)$.

Shifted Outcomes with an Exclusion Restriction

Empirical studies sometimes assume that y_1 and y_0 always differ by a constant, so that y_1 is a shifted version of y_0. For example, a long-standing concern of labor economics is to determine the effect of union membership on wages. Let y_1 be the wage that a person would earn if he or she were a union member and let y_0 be the wage that person would earn as a nonmember. Labor economists have often assumed that the *union wage differential* $y_1 - y_0$ is the same for all people.

Formally, assume there exists a constant k such that

$$(2.26) \quad y_1 = y_0 + k.$$

This assumption implies that, for all cutoff points t,

$$(2.27) \quad P(y_1 \le t \mid x, z = 0) = P(y_0 \le t - k \mid x, z = 0).$$

The distribution $P(y_0 \mid x, z = 0)$ is identified by the switching process. So $P(y_1 \mid x, z = 0)$ is identified if the value of the shift parameter k can be determined. It can be shown that k is identified if $P(y_1 \mid x)$ satisfies an exclusion restriction of the type discussed in Section 2.4.[11] Hence $P(y_1 \mid x)$ is identified if a switching process generates the data, outcomes are shifted, and an exclusion restriction holds.

This is a powerful finding, identifying $P(y_1 \mid x)$ without imposing any assumptions on the switching rule determining whether y_1 or y_0 is observed. The result is achieved at high cost, however. An exclusion restriction may or may not be available in a given application, but the assumption of shifted outcomes strains credibility.

Consider, for example, the union wage differential. Is it plausible to assume that union membership offers the same wage increment for all workers? Union contracts are often thought to tie wages and job security more closely to seniority than to merit. It therefore seems likely that, within a given job category, the less productive workers experience a larger union wage differential than do the more productive ones.

Ordered Outcomes

Similar in spirit to shifted outcomes is the assumption that y_1 and y_0 are ordered with, say, the value of y_1 always at least as large as the value of y_0. For example, suppose that an ill person may be treated by drug therapy ($z = 1$) or by placebo ($z = 0$). Let the outcomes y_1 and y_0 be the life span following each treatment. One might not know the value of drug therapy but might feel confident that it can do no harm. If so, then one can assume that y_1 must be at least as large as y_0.

Formally, the assumption is

$$(2.28) \quad y_1 \geq y_0.$$

This assumption implies that, for all cutoff points t,

$$(2.29) \quad P(y_1 \leq t \mid x, z = 0) \leq P(y_0 \leq t \mid x, z = 0).$$

Hence

$$
\begin{aligned}
(2.30) \quad P(y_1 \le t \mid x) &= P(y_1 \le t \mid x, z = 1) P(z = 1 \mid x) \\
&\quad + P(y_1 \le t \mid x, z = 0) P(z = 0 \mid x) \\
&\le P(y_1 \le t \mid x, z = 1) P(z = 1 \mid x) \\
&\quad + P(y_0 \le t \mid x, z = 0) P(z = 0 \mid x) \\
&= P(y \le t \mid x).
\end{aligned}
$$

On the one hand, the upper bound on $P(y_1 \le t \mid x)$ given here improves on the one in equation (2.7), which is the best available under censored sampling. On the other hand, the ordered-outcomes assumption does not tighten the lower bound in (2.7).

Selection of the Treatment with the Larger / Smaller Outcome

The shifted-outcomes and ordered-outcomes assumptions link the outcomes (y_1, y_0) by restricting the form of their joint distribution $P(y_1, y_0 \mid x)$. These outcomes may also be linked by assumptions on the switching rule determining whether y_1 or y_0 is realized. I shall discuss two symmetric cases.

Economic analyses of voluntary treatment policies often assume that the treatment yielding the larger outcome is selected. An example is the Roy (1951) model of occupation choice, where a person selects between two occupations by choosing the occupation with the higher wage. Let $z = 1$ if one occupation is chosen, $z = 0$ if the other, and let y_1 and y_0 be the wages that would be earned in the two occupations. Then the Roy model asserts that

$$
(2.31) \quad z = 1 \text{ if } y_1 > y_0 \quad \text{and} \quad z = 0 \text{ if } y_1 < y_0.
$$

No prediction is made if $y_1 = y_0$, but it is conventionally assumed that this event occurs with probability zero in the population. So the borderline case may be ignored.

To a researcher studying occupational wage determination, the Roy model is prior information specifying the switching rule determining whether y_1 or y_0 is observed. The situation is the same as in the reservation-wage model except that there, $z = 0$ implied that no outcome was observed, while here the outcome y_0 is observed.

Observability of y_0 makes a crucial difference. Whereas the reservation-wage model was earlier shown to have no identifying power in the absence of distributional assumptions (see Section 2.4), the Roy model does have identifying power. In particular, equation (2.31) implies that for all cutoff points t,

$$(2.32) \quad P(y_1 \leq t \mid x, z = 0) \geq P(y_0 \leq t \mid x, z = 0).$$

It follows that

$$
\begin{aligned}
(2.33) \quad P(y_1 \leq t \mid x) &= P(y_1 \leq t \mid x, z = 1)\, P(z = 1 \mid x) \\
&\quad + P(y_1 \leq t \mid x, z = 0)\, P(z = 0 \mid x) \\
&\geq P(y_1 \leq t \mid x, z = 1)\, P(z = 1 \mid x) \\
&\quad + P(y_0 \leq t \mid x, z = 0)\, P(z = 0 \mid x) \\
&= P(y \leq t \mid x).
\end{aligned}
$$

This lower bound on $P(y_1 \leq t \mid x)$ improves on the one in (2.7), which is the best available under censored sampling. The upper bound on $P(y_1 \leq t \mid x)$, however, remains as in (2.7).

The *competing-risks* model of survival analysis assumes that the treatment with the smaller outcome is selected, rather than the larger (see Kalbfleisch and Prentice, 1980). A person with two terminal diseases, for example, dies of the disease that first manifests itself.[12] Let $z = 1$ if one disease causes death, $z = 0$ if the other, and let y_1 and y_0 be the span of time before each disease manifests itself. Then the competing-risks model asserts that

$$(2.34) \quad z = 1 \text{ if } y_1 < y_0 \quad \text{and} \quad z = 0 \text{ if } y_1 > y_0.$$

It follows from (2.34) that

$$(2.35) \quad P(y_1 \leq t \mid x, z = 0) \leq P(y_0 \leq t \mid x, z = 0)$$

for all cutoff points t. This is the same finding as was reported in equation (2.29) under the ordered-outcomes assumption. Again (2.30) gives the upper bound on $P(y_1 \leq t \mid x)$.

2.7. Predicting High School Graduation If All Families Were Intact

To illustrate inference on treatment effects, I shall carry further the analysis of high school graduation begun in Section 1.5. The discussion there concluded by observing that the estimated graduation probabilities presented in Table 1.1 do not suffice to predict outcomes if the American environment of the 1980s were to change in some way. Among other things, these estimates do not permit one to predict what would happen if the fraction of intact families in the population were increased.

The study by Manski et al. (1992) goes beyond the estimates in Table 1.1 and addresses the extreme version of this question, namely: What would high school graduation probabilities be if all families were intact? The study also compares this scenario with the opposite extreme in which all families are nonintact.

We imagine that each child is characterized by two hypothetical high school graduation outcomes, y_1 and y_0. Variable y_1 indicates the outcome if the child were to reside in an intact family, with $y_1 = 1$ if the child would graduate and $y_1 = 0$ otherwise. Analogously, y_0 indicates the outcome if the same child were to reside in a nonintact family. We wish to learn the probability $P(y_1 = 1 \mid x)$ that a child with covariates x would graduate if all such children were to reside in intact families, and to compare $P(y_1 = 1 \mid x)$ with the probability $P(y_0 = 1 \mid x)$ of graduation if all children with covariates x were to reside in nonintact families. The covariates x are race, sex, and parents' schooling. In Section 1.5 family structure was also a covariate, but this variable now appears as a treatment instead.

The inferential problem stems from the fact that each child in the NLSY sample actually realizes only one of the two graduation outcomes; y_1 is realized if a child actually resides in an intact family ($z = 1$) and y_0 is realized otherwise ($z = 0$). The empirical evidence therefore reveals the probability $P(z = 1 \mid x)$ that a child with covari-

Table 2.1 Estimated probabilities of residence in an intact family, for children
whose parents have completed twelve years of schooling

White male	White female	Black male	Black female
.82	.82	.54	.46

Source: Computations based on Manski et al. (1992), table 4.

ates x resides in an intact family, the probability $P(y_1 = 1 \mid x, z = 1)$ of graduation conditional on residing in an intact family, and the probability $P(y_0 = 1 \mid x, z = 0)$ of graduation conditional on residing in a nonintact family.

As in Section 1.5, I focus here on children whose parents have both completed twelve years of schooling. Estimates of $P(y_1 = 1 \mid x, z = 1)$ and $P(y_0 = 1 \mid x, z = 0)$ have already been presented in Table 1.1. Estimates of $P(z = 1 \mid x)$ are given in Table 2.1. The striking feature of the table is the difference across races. Holding sex and parents' schooling fixed, we find that a white child is much more likely than a black child to reside in an intact family at age 14.

The data in Tables 1.1 and 2.1 provide all the components needed to infer the graduation probabilities $P(y_1 = 1 \mid x)$ and $P(y_0 = 1 \mid x)$ under the various assumptions considered in Sections 2.5 and 2.6. Table 2.2 presents the estimates for $P(y_1 = 1 \mid x)$. Similar estimates may be computed for $P(y_0 = 1 \mid x)$.

Table 2.2 Estimated bounds on high school graduation probabilities if all families were intact, for children whose parents have completed twelve years of schooling

Prior information	White male	White female	Black male	Black female
No prior information (equation 2.6)	[.73, .91]	[.77, .95]	[.47, .93]	[.44, .98]
Ordered outcomes (equation 2.30)	[.87, .91]	[.91, .95]	[.83, .93]	[.91, .98]
Treatment with the larger outcome (equation 2.33)	[.73, .87]	[.77, .91]	[.47, .83]	[.44, .91]
Treatment independent of outcomes (equation 2.24)	.89	.94	.87	.95

Source: Computations based on Tables 1.1 and 2.1.

Table 2.2 makes clear how prior information affects the conclusions one may draw about graduation outcomes if all families were intact. In the absence of prior information, one may estimate the bounds in the top row of the table. The width of each bound is the estimated probability of residing in a nonintact family.

Prior information may allow one to tighten the worst-case bounds. Suppose one believes that residing in an intact family can never harm a child's schooling prospects; so outcomes are ordered with $y_1 \geq y_0$ for all children. Then mandating that all families be intact cannot decrease graduation probabilities relative to those actually found among the NLSY respondents. The actual graduation probabilities, namely,

$$(2.36) \quad P(y = 1 \mid x) = P(y_1 = 1 \mid x, z = 1) P(z = 1 \mid x)$$
$$+ P(y_0 = 1 \mid x, z = 0) P(z = 0 \mid x),$$

are estimated to be .87 for white males, .91 for white females, .83 for black males, and .91 for black females. These values become lower bounds under the ordered-outcomes assumption.

Suppose one believes that realized family structure reflects parents' decisions about what is best for their children's schooling prospects—in other words, that families choose the treatment with the larger outcome. (For example, parents may compare the likely impact on their children of maintaining a marriage characterized by constant fighting and hostility with the impact of raising the children in a single-parent household.) If family structure is determined in this manner, then mandating that all families be intact cannot increase graduation probabilities relative to those actually found among the NLSY respondents, and may decrease them. So the actual graduation probabilities now become upper bounds on $P(y_1 = 1 \mid x)$ rather than lower bounds.

Finally, suppose one believes that family structure is exogenous with respect to children's schooling prospects. Then the graduation probabilities $P(y_1 = 1 \mid x)$ coincide with the probabilities $P(y_1 = 1 \mid x, z = 1)$ reported earlier in the left column of Table 1.1, and now found in the fourth row of Table 2.2.

Each of the rows of Table 2.2 presents a logically valid conclusion about the graduation outcomes that would be realized if all families were intact. Any social scientist who accepts the NLSY as empirical evidence must agree that the probabilities $P(y_1 = 1 \mid x)$ lie within the worst-case bounds of the first row. Beyond that, the conclusions one draws necessarily depend on the assumptions one is willing to maintain.

3

The Mixing Problem in Program Evaluation

3.1. The Experimental Evaluation of Social Programs

In the United States, concern with the evaluation of social programs has spread rapidly since the 1960s, when attempts were made to evaluate the impacts of programs proposed as part of the War on Poverty. Evaluation requirements now appear in major federal statutes. One of these is the Family Support Act of 1988, which revised the AFDC program. In Title II of this statute, Congress mandated study of the effectiveness of training programs initiated by the states under the new Job Opportunities and Basic Skills Training Program (JOBS). Congress even stipulated the mode of data collection: "a demonstration project conducted under this subparagraph shall use experimental and control groups that are composed of a random sample of participants in the program" (Public Law 100–485, October 13, 1988, section 203, 102 stat. 2380).

Until the early 1980s, evaluations of welfare and training programs generally analyzed data from ongoing programs rather than from controlled social experiments.[1] Some studies analyzed cross-sectional variation in outcomes across cities or states that had different programs. Others analyzed "before-and-after" data; that is, time-series variation within a city or state that altered its program.

More recently, social experiments have come to dominate the evaluations commissioned by the federal government and by the major foundations. Dissatisfaction with the evaluations of job training programs performed in the 1970s led the U.S. Department of Labor to commission an experimental evaluation of the Job Training Partnership Act in the mid-1980s (see Hotz, 1992). And a set of experiments sponsored by the Ford Foundation and executed by the Man-power Demonstration Research Corporation influenced the federal government to choose experimental analysis as the preferred approach to evaluations of AFDC reforms (see Greenberg and Wiseman, 1992).

Controlled social experimentation has become so much the new orthodoxy of evaluation that Jo Anne Barnhart, an Assistant Secretary in the U.S. Department of Health and Human Services during the Bush administration, could write this about the evaluation of training programs for welfare recipients: "In fact, nonexperimental research of training programs has shown such methods to be so unreliable, that Congress and the Administration have both insisted on experimental designs for the Job Training Partnership Act (JTPA) and the Job Opportunities and Basic Skills (JOBS) programs" (letter from Jo Anne B. Barnhart to Eleanor Chelimsky, reproduced as U.S. General Accounting Office, 1992, appendix II).

Barnhart's reference to the unreliability of nonexperimental research reflects the view of some social scientists that the selection problem analyzed in Chapter 2 precludes credible inference on treatment effects from the observation of outcomes when treatments are uncontrolled. These social scientists have recommended that empirical research should focus exclusively on the design and analysis of controlled experiments. See Bassi and Ashenfelter (1986), LaLonde (1986), and Coyle, Boruch, and Turner (1989).

The Classical Argument

Recent arguments for controlled social experiments follow the wonderfully simple, classical lines of Fisher (1935). In a controlled experiment, random samples of persons with specified covariates are drawn and formed into treatment groups. All members of a treatment group are assigned the same treatment. The empirical distribution of outcomes realized by a treatment group is then ostensibly the same (up to random sampling error) as would be observed if the treatment in question were applied to all persons with the specified covariates.

Formally, recall the setup of Section 2.5. There are two mutually exclusive treatments, labeled 1 and 0. Each member of the population is described by values for the variables (y_1, y_0, z, x). The outcome y_1 is observed if a person receives treatment 1, y_0 is observed if the person receives treatment 0, and z indicates which treatment is received. The sampling process identifies $P(y_1 \mid x, z = 1)$ and $P(y_0 \mid x, z = 0)$. Randomized selection of treatment implies that treatment is statistically independent of the outcomes. Hence $P(y_1 \mid x, z = 1) = P(y_1 \mid x)$ and $P(y_0 \mid x, z = 0) = P(y_0 \mid x)$. A controlled experiment thus reveals $P(y_1 \mid x)$ and $P(y_0 \mid x)$.

Extrapolation from Social Experiments

The classical argument for experimentation does not suffice to conclude that experiments may be used to predict the outcomes of social programs. Experimental evaluation also requires a critical invariance assumption: the experimental version of a program must operate as would an actual program. It is this premise that allows one to extrapolate from the experiment to the real world.

Critiques of social experimentation argue that, for a variety of reasons, experimental versions of social programs may not operate as would actual programs (see Hausman and Wise, 1985, and Manski and Garfinkel, 1992). Four widely recognized concerns are described here. A fifth, previously unexplored, issue will be examined in depth beginning in Section 3.2.

> *Program administration.* Experiments with randomized treatments may be administered differently from actual programs mandating homogeneous treatment of the population. Social experiments generally cannot be performed using the double-blind protocols of medical trials, in which neither experimenters nor subjects know who is in each treatment group. Program administrators inevitably know who is in each group and cannot be prevented from using this information to influence outcomes.

> *Macro feedback effects.* Full-scale programs may change the environment in ways that influence outcomes. Possible feedbacks range from labor market equilibration to information diffusion to norm formation. The scale of the typical social experiment is too small to discern these macro effects, which may become prominent when a program is actually implemented (see Garfinkel, Manski, and Michalopolous, 1992).

> *Site selection.* The classical experimental paradigm calls for random selection of treatment sites, but evaluators generally do not

have the power to compel sites to cooperate. Hence experiments are typically conducted at sites selected jointly by evaluators and local officials. Hotz (1992) describes how the JTPA evaluators originally sought to select sites randomly but, being unable to secure the agreement of the randomly drawn sites, were ultimately required to provide large financial incentives to nonrandomly chosen sites in order to obtain their cooperation.

Program participation. A classical social experiment randomly assigns the participants in a social program to groups receiving different treatments. Analyis of the experimental data presumes that the existence of the experiment does not alter the population of persons participating in the program. Heckman (1992) and Moffitt (1992b) observe that it is not plausible to assume an invariant population of participants, because the value to a person of participating in a program with randomized treatment is not the same as that of participating in a program with known treatment. Heckman makes the point well when he observes that social experimentation is intrinsically different from experimentation in the biological sciences and agriculture: "Plots of ground do not respond to anticipated treatments of fertilizer, nor can they excuse themselves from being treated" (p. 215).

None of the foregoing concerns implies that experimental evidence is uninformative. These concerns do imply that one should not expect the distribution of outcomes in a randomly selected treatment group to coincide with the outcomes that would be realized in an actual social program. Extrapolation from experimental data, as from nonexperimental data, requires that the empirical evidence be combined with prior information.

3.2. Variation in Treatment

Let us abstract from the very real concerns just expressed and accept the classical argument that the outcomes realized by a randomly selected treatment group are the same (up to random sampling error) as would be observed if the treatment were applied to all persons with specified covariates. Policies mandating homogeneous treatment of the population are of interest, but so are ones that permit treatment to vary across the population. We often see policies calling on persons to select their own treatments. Policies intended to mandate homogeneous treatment sometimes turn out to be voluntary in practice, because compliance with the mandated treatment is not enforced. Resource constraints sometimes prevent universal implementation of desirable treatments.

Consider the following inferential questions:

What do observations of outcomes when treatments vary across the population reveal about the outcomes that would occur if treatment were homogeneous?

What do observations of outcomes when treatment is homogeneous reveal about the outcomes that would occur if treatment were to vary across the population?

The first question poses the selection problem. The second question, which has remained unexplored and unnamed, is the subject of this chapter. Formally, the question asks what inferences about mixtures of two random variables can be made given knowledge of their marginal distributions. Hence I refer to it as the *mixing* problem.[2]

To state the mixing problem formally, let us enhance slightly the description of the population given in Section 2.5 by making explicit the *treatment policy,* or assignment rule, being implemented. Let each member of the population be described by values for $[(y_1, y_0), z_m, x]$. As before, there are two mutually exclusive treatments, labeled 1 and 0, and (y_1, y_0) are the outcomes associated with the two treatments.

A treatment policy, now denoted m, determines which treatment each person receives. The indicator variable z_m denotes the treatment received under policy m; $z_m = 1$ if the person receives treatment 1, and $z_m = 0$ otherwise. The outcome a person realizes under policy m is y_1 if $z_m = 1$ and y_0 otherwise. Let y_m denote the realized outcome; that is,

$$(3.1) \qquad y_m \equiv y_1 z_m + y_0(1 - z_m).$$

The distribution of outcomes realized by persons with covariates x is

$$(3.2) \quad P(y_m \mid x) \equiv P[y_1 z_m + y_0(1 - z_m) \mid x]$$
$$= P(y_1 \mid x, z_m = 1) P(z_m = 1 \mid x)$$
$$+ P(y_0 \mid x, z_m = 0) P(z_m = 0 \mid x).$$

A welfare recipient, for example, might be treated by job-specific training or by basic education. The relevant outcome might be earned income following treatment. One treatment policy might mandate the job training treatment for all welfare recipients and enforce the mandate. A second policy might attempt to mandate the basic education treatment but not be able to enforce compliance. A third policy might permit a person's caseworker to select the treatment expected to yield the larger net benefit, measured as earned income minus treatment costs.

The problem of interest is to learn about the distribution $P(y_m \mid x)$ of outcomes that would be realized by persons with covariates x if a specified treatment policy m were in effect. Inference is straightforward if one can enact policy m and observe the outcomes. The interesting inferential questions concern the feasibility of learning $P(y_m \mid x)$ when one observes outcomes under policies other than m. The selection problem and the mixing problem both concern the feasibility of extrapolating from observed treatment policies to unobserved ones.

The selection problem arises when policy m mandates homogeneous treatment, but the available data are realizations under some other policy that may yield heterogeneous treatments. Suppose that m makes treatment 1 mandatory for all persons with covariates x, so $P(z_m = 1 \mid x) = 1$ and $P(y_m \mid x) = P(y_1 \mid x)$. Suppose that the observable policy is some $\mu \neq m$.[3] The sampling process identifies the censored outcome distributions $P(y_1 \mid x, z_\mu = 1)$ and $P(y_0 \mid x, z_\mu = 0)$, as well as the treatment distribution $P(z_\mu \mid x)$. Then the formal statement of the selection problem is:

What does knowledge of $[P(y_1 \mid x, z_\mu = 1), P(y_0 \mid x, z_\mu = 0), P(z_\mu \mid x)]$ imply about $P(y_1 \mid x)$?

The mixing problem arises when policy m may yield heterogeneous treatments, but the available data are realizations under policies imposing homogeneous treatments. In particular, the classical model of experimentation presumes that experimental evidence is available for both treatments, so the experiments identify $P(y_1 \mid x)$ and $P(y_0 \mid x)$. Therefore the formal statement of the mixing problem is:[4]

What does knowledge of $[P(y_1 \mid x),\ P(y_0 \mid x)]$ imply about $P[y_1 z_m + y_0(1 - z_m) \mid x]$?

Section 3.3 uses empirical evidence from a famous social experiment, the Perry Preschool Project, to illustrate the mixing problem and the main findings of the chapter. Fifteen years after their participation in an early childhood educational intervention, 67 percent of an experimental group were high school graduates. At the same time, only 49 percent of a control group were graduates. Our interest is to determine what the experimental evidence and various assumptions imply about the rate of high school graduation that would prevail under treatment policies applying the intervention to some children but not to others.

Sections 3.4 through 3.7 present the analysis that yields the empirical results reported in Section 3.3. In studying the selection problem in Chapter 2, we found it productive to begin by determining what can be learned when the sampling process provides the only information available to the researcher. We then examined the identifying power of various forms of prior information that might plausibly be invoked in empirical studies. The present analysis uses the same approach.

Section 3.4 investigates the mixing problem when knowledge of the two marginal distributions $P(y_1 \mid x)$ and $P(y_0 \mid x)$ is the only information available. Then Sections 3.5 through 3.7 explore the identifying power of assumptions that restrict the determinants of outcomes.[5] Section 3.5 examines the implications of assumptions restricting the joint distribution of the outcomes (y_1, y_0). Section 3.6 examines assumptions restricting the treatment policy. Section 3.7 cites some combinations of assumptions that identify $P(y_m \mid x)$.

The mixing problem, like the selection problem, is a failure of identification rather than a difficulty in sample inference. To keep attention focused on identification, Sections 3.4 through 3.7 maintain the assumption that the distributions $[P(y_1 \mid x), P(y_0 \mid x)]$ are known. The identification findings reported in these sections can be translated into consistent sample estimates of identified quantities by replacing $P(y_1 \mid x)$ and $P(y_0 \mid x)$ with consistent estimates, as is done in Section 3.3.

3.3. The Perry Preschool Project

Beginning in 1962, the Perry Preschool Project provided intensive educational and social services to a random sample of disadvantaged black children in Ypsilanti, Michigan. The project investigators also drew a second random sample of such children but provided them with no special services. Subsequently, a variety of outcomes were ascertained for most members of the experimental and control groups. Among other things, it was found that 67 percent of the experimental group and 49 percent of the control group were high school graduates by age 19 (see Berrueta-Clement et al., 1984). This and similar findings for other outcomes have been widely cited as evidence that intensive early childhood educational interventions improve the outcomes of children from disadvantaged backgrounds (see Holden, 1990).

For purposes of discussion, let us accept the Perry Preschool Project as a classical controlled experiment, with

x = black children in Ypsilanti, Michigan,
y_1 = outcome if a child were to be assigned to the experimental group ($y_1 = 1$ if high school graduate by age 19, $= 0$ otherwise), and
y_0 = outcome if a child were to be assigned to the control group ($y_0 = 1$ if high school graduate by age 19, $= 0$ otherwise).

Moreover, ignoring attrition and sampling error in the estimation of outcome distributions, let us accept the experimental evidence as showing that the high school graduation probability among children with covariates x would be .67 if all such children were to receive the intervention, and .49 if none of them was to receive the intervention. That is, let us accept the experimental evidence as showing that $P(y_1 = 1 \mid x) = .67$ and $P(y_0 = 1 \mid x) = .49$.[6]

What would be the probability of high school graduation under a treatment policy in which some children with covariates x receive the intervention, but not others? Table 3.1 summarizes the inferences that can be made given the experimental evidence and varying forms of prior information about the outcome distribution and the treatment policy. The remainder of this section discusses the empirical findings. Sections 3.4 through 3.7 present the analysis underlying these findings.

Identification Using Only the Experimental Evidence

It might be conjectured that $P(y_m = 1 \mid x)$ must lie between the graduation probabilities of the control and treatment groups, namely, between .49 and .67. This conjecture is correct for special outcome distributions and treatment policies. It holds if (a) the outcomes (y_1, y_0)

Table 3.1   The Perry Preschool Project

Experimental Evidence

$P(y_1 = 1 \mid x) = .67$ $P(y_0 = 1 \mid x) = .49$

Prior Information	$P(y_m = 1 \mid x)$
No prior information	[.16, 1]
Independent outcomes	[.33, .83]
Ordered outcomes	[.49, .67]
Treatment independent of outcomes	[.49, .67]
Treatment maximizing graduation probability	[.67, 1]
+ independent outcomes	.83
+ ordered outcomes	.67
Treatment minimizing graduation probability	[.16, .49]
1/10 population receives treatment 1	[.39, .59]
+ treatment independent of outcomes	.51
5/10 population receives treatment 1	[.17, .99]
+ treatment independent of outcomes	.58
9/10 population receives treatment 1	[.57, .77]
+ treatment independent of outcomes	.65

Source: Manski (1994b), table 1.

are ordered, with $y_1 \geq y_0$ for all children, or if (b) the treatment policy makes z_m statistically independent of the outcomes (y_1, y_0).

The conjecture does not hold more generally. In fact, the experimental evidence only implies that the graduation probability must lie between .16 and 1. That is, there exist outcome distributions and treatment policies that are consistent with the known values of $P(y_1 \mid x)$ and $P(y_0 \mid x)$ and that imply graduation probabilities as low as .16 and as high as 1.

This result is easily understood once one considers precisely what the experimental evidence reveals. Observing the outcomes of the treatment group reveals (ignoring sampling error) that $y_1 = 1$ for 67 percent of the population and $y_1 = 0$ for the remaining 33 percent. Observing the outcomes of the control group reveals that $y_0 = 1$ for 49 percent of the population and $y_0 = 0$ for the remaining 51 percent.

The experimental evidence does not reveal how y_1 and y_0 are related within the population, nor how policy m assigns treatments. The impact of treatment policy on the graduation rate is most pronounced when y_1 and y_0 are most negatively associated. Among all distributions of (y_1, y_0) that are consistent with the experimental evidence, the one with the greatest negative association between y_1 and y_0 is this:

$$P(y_1 = 0, y_0 = 0 \mid x) = .00 \qquad P(y_1 = 0, y_0 = 1 \mid x) = .33$$

$$P(y_1 = 1, y_0 = 0 \mid x) = .51 \qquad P(y_1 = 1, y_0 = 1 \mid x) = .16.$$

Given this distribution of outcomes, the graduation rate is maximized by adopting a treatment policy that provides the intervention only to those children with $y_1 = 1$. The result is a 100 percent graduation rate. At the other extreme, the graduation probability is minimized by adopting a treatment policy that provides the intervention only to those children with $y_1 = 0$. The result is a 16 percent graduation rate.

Prior Information

The interval $[.16, 1]$ is a worst-case bound on the graduation probability, computed in the absence of any prior information restricting the outcome distribution or the treatment policy. A researcher who possesses such information may be able to narrow the range of possible graduation probabilities.

Imagine that one has no information about the treatment policy but does have information about the outcome distribution. One might think that being treated by the preschool intervention can never harm a child's schooling prospects; that is, outcomes are ordered with $y_1 \geq y_0$ for all children. If so, then the graduation probability must lie between those observed in the control and treatment groups, namely, between .49 and .67. A more neutral assumption might be that y_1 and y_0 are statistically independent conditional on x. This assumption implies that the graduation probability must lie between .33 and .83; where the probability falls within this range depends on the treatment policy.

Next imagine that one has no information about the outcome distribution but does have information about the treatment policy. One might think that treatment decisions will be made by omniscient parents who choose for each child the treatment yielding the better outcome. This assumption implies that the graduation rate must lie between .67 and 1; where the rate falls within this range depends on the outcome distribution. Or one might think that assignments to treatments are statistically independent of outcomes, as they would be if an explicit random assignment rule were used. Then the graduation rate must lie between the .49 and .67 observed in the control and treatment groups.

Finally, imagine that resource constraints limit implementation of the intervention to part of the population. Suppose that one knows the fraction of the population receiving the intervention, but does not know the composition of the treated and untreated subpopulations. As Table 3.1 shows, knowing that $1/10$ or $5/10$ or $9/10$ of the population receives the intervention implies that the graduation rate must lie in the interval [.39, .59] or [.17, .99] or [.57, .77], respectively. Observe that the first and third intervals are relatively narrow but the second is rather wide, almost as wide as the interval found in the absence of prior information. This pattern of results reflects the fact that the power of treatment policy to determine who receives which treatment is much more constrained when $P(z_m = 1 \mid x)$ is fixed at a value near zero or one than it is when $P(z_m = 1 \mid x)$ is fixed at $5/10$.

The scenarios considered thus far bring to bear enough empirical evidence and prior information to bound the high school graduation rate but not to identify it. If stronger restrictions are imposed, then the high school graduation rate may be identified. For example, suppose it is known that the outcomes y_1 and y_0 are statistically independent of one another and that each child receives the treatment yielding the better outcome. Then the implied high school graduation rate is .83. Or suppose it is known that $5/10$ of the population receives the intervention and that the treatment z is independent of the outcomes (y_1, y_0), as defined in Section 2.5. Then the implied graduation rate is .58.

The general lesson is that experimental evidence alone permits only weak conclusions to be drawn about the high school graduation rate when treatments vary. Experimental evidence combined with prior information implies stronger conclusions. The nature of these stronger conclusions depends critically on the prior information asserted. This lesson is analogous to the one learned over the past twenty years about the conclusions that can be drawn about mandatory programs from observations of outcomes when treatments vary. Mixing and selection are distinct identification problems, but they are closely related.

3.4. Identification of Mixtures Using Only Knowledge of the Marginals

This section characterizes the restrictions on $P(y_m \mid x)$ implied by knowledge of $[P(y_1 \mid x), P(y_0 \mid x)]$ in the worst-case situation where no other information is available. Here and throughout the remainder of the chapter, I focus on the basic problem of inferring conditional probabilities of events. Manski (1994b) shows how these findings can be used to study the identification of quantiles of $P(y_m \mid x)$.

Consider the probability $P(y_m \in B \mid x)$ that the realized outcome y_m falls in a specified set B, conditional on x. Given that y_m always equals either y_1 or y_0, one might think that $P(y_m \in B \mid x)$ must lie between $P(y_1 \in B \mid x)$ and $P(y_0 \in B \mid x)$. This is not the case. It turns out that when $P(y_1 \in B \mid x) + P(y_0 \in B \mid x) \leq 1$, then $P(y_m \in B \mid x)$ must lie in the interval $[0, P(y_1 \in B \mid x) + P(y_0 \in B \mid x)]$. When

$P(y_1 \in B \mid x) + P(y_0 \in B \mid x) \geq 1$, $P(y_m \in B \mid x)$ must lie in the interval $[P(y_1 \in B \mid x) + P(y_0 \in B \mid x) - 1, 1]$. That is, when $P(y_1 \mid x)$ and $P(y_0 \mid x)$ are known but no information restricting the distribution of $[(y_1, y_0), z_m, x]$ is available, we can conclude only that

$$(3.3) \quad \max[0, P(y_1 \in B \mid x) + P(y_0 \in B \mid x) - 1] \leq P(y_m \in B \mid x)$$
$$\leq \min[P(y_1 \in B \mid x) + P(y_0 \in B \mid x), 1].$$

This central finding may appear unintuitive, so I shall prove it here. We first determine the treatment policies that minimize and maximize $P(y_m \in B \mid x)$. Observe that if y_1 and y_0 both fall in the set B, then y_m must fall in B. Moreover, if neither y_1 nor y_0 falls in B, then y_m cannot fall in B. That is,

$$(3.4a) \quad y_1 \in B \cap y_0 \in B \Rightarrow y_m \in B$$

and

$$(3.4b) \quad y_1 \notin B \cap y_0 \notin B \Rightarrow y_m \notin B,$$

whatever treatment policy m may be.

The treatment policy is relevant in those cases in which one of the two outcomes falls in B and the other does not. The treatment policy minimizes $P(y_m \in B \mid x)$ if it always selects the treatment yielding the outcome that is not in B; that is, if

$$(3.5) \quad y_1 \notin B \cap y_0 \in B \Rightarrow z_m = 1$$
$$y_1 \in B \cap y_0 \notin B \Rightarrow z_m = 0.$$

It follows that the smallest possible value of $P(y_m \in B \mid x)$ is $P(y_1 \in B \cap y_0 \in B \mid x)$. The treatment policy maximizes $P(y_m \in B \mid x)$ if it always selects the treatment yielding the outcome that is in B; that is, if

$$(3.6) \quad y_1 \notin B \cap y_0 \in B \Rightarrow z_m = 0$$
$$y_1 \in B \cap y_0 \notin B \Rightarrow z_m = 1.$$

Therefore the largest possible value of $P(y_m \in B \mid x)$ is $P(y_1 \in B \cup y_0 \in B \mid x)$.

The above shows that if both $P(y_1 \in B \cap y_0 \in B \mid x)$ and $P(y_1 \in B \cup y_0 \in B \mid x)$ are known, then

(3.7) $P(y_1 \in B \cap y_0 \in B \mid x) \leq P(y_m \in B \mid x)$

$\leq P(y_1 \in B \cup y_0 \in B \mid x)$

is the sharp bound on $P(y_m \in B \mid x)$. But the only available information is knowledge of $P(y_1 \mid x)$ and $P(y_0 \mid x)$. Therefore the best computable lower bound on $P(y_m \in B \mid x)$ is the smallest value of $P(y_1 \in B \cap y_0 \in B \mid x)$ that is consistent with the known $P(y_1 \mid x)$ and $P(y_0 \mid x)$. Similarly, the best computable upper bound is the largest feasible value of $P(y_1 \in B \cup y_0 \in B \mid x)$.

The second step is to determine these best computable bounds. This is simple to do, because Frechet (1951) proved this sharp bound on $P(y_1 \in B \cap y_0 \in B \mid x)$:[7]

(3.8) $\max[0, P(y_1 \in B \mid x) + P(y_0 \in B \mid x) - 1]$

$\leq P(y_1 \in B \cap y_0 \in B \mid x)$

$\leq \min[P(y_1 \in B \mid x), P(y_0 \in B \mid x)]$.

It follows immediately from (3.8) that the best computable lower bound on $P(y_m \mid x)$ is $\max[0, P(y_1 \in B \mid x) + P(y_0 \in B \mid x) - 1]$. To obtain the best computable upper bound, observe that

(3.9) $P(y_1 \in B \cup y_0 \in B \mid x) = P(y_1 \in B \mid x) + P(y_0 \in B \mid x)$

$- P(y_1 \in B \cap y_0 \in B \mid x)$.

Applying the Frechet lower bound on $P(y_1 \in B \cap y_0 \in B \mid x)$ to (3.9) shows that

(3.10) $P(y_1 \in B \cup y_0 \in B \mid x) \leq \min[P(y_1 \in B \mid x) + P(y_0 \in B \mid x), 1]$.

Hence $\min[P(y_1 \in B \mid x) + P(y_0 \in B \mid x), 1]$ is the best computable upper bound.

3.5. Restrictions on the Outcome Distribution

In the preceding section, we showed that if $P(y_1 \in B \cap y_0 \in B \mid x)$ and $P(y_1 \in B \cup y_0 \in B \mid x)$ are known and if no restrictions are imposed on the treatment policy m, then inequality (3.7) provides the sharp bound on $P(y_m \in B \mid x)$. One may sometimes have prior information that makes the bound (3.7) computable. This section presents three cases.

Independent Outcomes

Suppose it is known that outcomes y_1 and y_0 are statistically independent, conditional on x. Then

$$(3.11) \quad P(y_1 \in B \cap y_0 \in B \mid x) = P(y_1 \in B \mid x)\,P(y_0 \in B \mid x).$$

With $P(y_1 \in B \cap y_0 \in B \mid x)$ known, the sharp bound on $P(y_m \in B \mid x)$ is

$$(3.12) \quad P(y_1 \in B \mid x)\,P(y_0 \in B \mid x) \le P(y_m \in B \mid x)$$
$$\le P(y_1 \in B \mid x) + P(y_0 \in B \mid x)$$
$$- P(y_1 \in B \mid x)\,P(y_0 \in B \mid x).$$

Whereas the worst-case bound obtained in Section 3.4 was informative only in one direction, the present bound is generally informative both above and below. The new lower bound on $P(y_m \mid x)$ is informative when $P(y_1 \in B \mid x)$ and $P(y_0 \in B \mid x)$ are positive. The upper bound is informative when $P(y_1 \in B \mid x)$ and $P(y_0 \in B \mid x)$ are both less than 1.

Shifted Outcomes

Evaluation studies often assume that y_1 and y_0 are not only statistically dependent but functionally dependent. It is especially common to assume that outcomes are shifted versions of one another; that is,[8]

$$(3.13) \quad y_1 = y_0 + k,$$

for some constant k. The assumption of shifted outcomes was discussed in Section 2.6.

Suppose it is known that (3.13) holds. Knowledge of $P(y_1 \mid x)$ and $P(y_0 \mid x)$ implies knowledge of k. So the joint distribution $P(y_1, y_0 \mid x)$ is known. With $P(y_1 \in B \cap y_0 \in B \mid x)$ known, the sharp bound on $P(y_m \in B \mid x)$ is

$$(3.14) \quad P[(y_0 + k) \in B \cap y_0 \in B \mid x] \leq P(y_m \in B \mid x)$$
$$\leq P[(y_0 + k) \in B \mid x] + P(y_0 \in B \mid x)$$
$$- P[(y_0 + k) \in B \cap y_0 \in B \mid x].$$

When B is the set of all real numbers below some cutoff point t, this bound takes a simple form. Assume, without loss of generality, that $k \geq 0$. Then (3.14) becomes

$$(3.15) \quad P(y_0 \leq t - k \mid x) \leq P(y_m \leq t \mid x) \leq P(y_0 \leq t \mid x)$$

or, equivalently,

$$(3.15') \quad P(y_1 \leq t \mid x) \leq P(y_m \leq t \mid x) \leq P(y_0 \leq t \mid x).$$

Ordered Outcomes

Outcomes y_1 and y_0 are said to be ordered with respect to a set B if y_0 always falls in B when y_1 does; that is,[9]

$$(3.16) \quad y_1 \in B \Rightarrow y_0 \in B.$$

For example, let the outcomes be binary, taking the value 0 or 1. If $y_1 = 0 \Rightarrow y_0 = 0$, then the outcomes are ordered with respect to the set $B = \{0\}$. Another example was given in Section 2.6, which considered the case

$$(3.17) \quad y_1 \geq y_0.$$

Here y_1 and y_0 are ordered with respect to all sets of the form $B = (-\infty, t]$, as $y_1 \leq t \Rightarrow y_0 \leq t$.

The assumption of ordered outcomes has earlier been discussed in the context of the Perry Preschool Project. One may believe that receiving the intervention cannot possibly diminish a child's prospects for high school graduation. If so, then any child who receives the intervention and does not graduate is a child who would not graduate in the absence of the intervention. That is, $y_1 = 0 \Rightarrow y_0 = 0$.

If y_1 and y_0 are ordered with respect to B, then

$$(3.18) \quad P(y_1 \in B \cap y_0 \in B \mid x) = P(y_1 \in B \mid x).$$

With $P(y_1 \in B \cap y_0 \in B \mid x)$ known, the sharp bound on $P(y_m \in B \mid x)$ is

$$(3.19) \quad P(y_1 \in B \mid x) \leq P(y_m \in B \mid x) \leq P(y_0 \in B \mid x).$$

An interesting result emerges when (3.19) is applied to outcomes satisfying (3.17). Letting $B = (-\infty, t]$, we find that (3.19) coincides with the bound (3.15′) that holds when outcomes are known to be shifted. Thus it turns out that, in the context of the mixing problem, assumptions (3.13) and (3.17) have the same power to identify $P(y_m \leq t \mid x)$. Section 2.6 showed that these two assumptions have different identifying power in the context of the selection problem.

3.6. Restrictions on the Treatment Policy

This section examines the restrictions on $P(y_m \mid x)$ implied by a set of polar treatment policies, in the absence of prior information about the outcome distribution. We first suppose that treatment is statistically independent of outcomes, as in random assignment policies. We then suppose that treatment minimizes or maximizes the probability that the realized outcome y_m falls in specified sets B, as in competing-risks models and in the Roy model. The section also examines the quite different problem of inference when the fraction of the population receiving each treatment is known, but nothing is known about the composition of the subpopulations receiving each treatment.

Treatment Independent of Outcomes

Suppose it is known that, under policy m, the treatment z_m received by each person is statistically independent of the person's outcomes (y_1, y_0). That is,

$$(3.20) \quad P[(y_1, y_0) \mid x, z_m] = P[(y_1, y_0) \mid x].$$

Then equation (3.2) reduces to

$$(3.21) \quad P(y_m \mid x) = P(y_1 \mid x) P(z_m = 1 \mid x) + P(y_0 \mid x) P(z_m = 0 \mid x).$$

If the fractions $P(z_m \mid x)$ of the population receiving each treatment are known, then $P(y_m \mid x)$ is identified. Our present concern, however, is with the situation in which (3.20) is the only prior information available. In this case, the only restriction on the treatment distribution is that $P(z_m = 1 \mid x)$ and $P(z_m = 0 \mid x)$ must lie in the unit interval and add up to one. Hence (3.21) implies that $P(y_m \in B \mid x)$ must lie between $P(y_1 \in B \mid x)$ and $P(y_0 \in B \mid x)$. That is,

$$(3.22) \quad \min[P(y_1 \in B \mid x), P(y_0 \in B \mid x)]$$
$$\leq P(y_m \in B \mid x) \leq \max[P(y_1 \in B \mid x), P(y_0 \in B \mid x)].$$

The bound (3.22) is contained within each of the bounds reported in Section 3.5, which left the treatment policy unspecified and imposed restrictions on the outcome distribution. This fact has a simple explanation. Equation (3.21) shows that, if treatment is independent of the outcomes, then $P(y_m \mid x)$ depends on the distribution of (y_1, y_0) only through the two marginal distributions $P(y_1 \mid x)$ and $P(y_0 \mid x)$.

Hence if one knows that treatment is independent of the outcomes, then restrictions on the joint distribution of (y_1, y_0) have no identifying power.

Optimizing Treatments

To derive the worst-case bound (3.3), we constructed two extreme treatment policies, one minimizing $P(y_m \in B \mid x)$ and the other maximizing it. The former policy satisfies equation (3.5), while the latter satisfies (3.6). Suppose that one of these optimizing policies is implemented. What can be learned about $P(y_m \in B \mid x)$ in the absence of prior restrictions on the outcome distribution?

The derivation of (3.3) showed that the treatment policy minimizing $P(y_m \in B \mid x)$ makes $P(y_m \in B \mid x) = P(y_1 \in B \cap y_0 \in B \mid x)$, while the policy maximizing $P(y_m \in B \mid x)$ makes $P(y_m \in B \mid x) = P(y_1 \in B \cup y_0 \in B \mid x)$. Applying the Frechet bound (3.8) shows that when the treatment policy is known to minimize $P(y_m \in B \mid x)$, we obtain the bound

$$(3.23) \quad \max[0, P(y_1 \in B \mid x) + P(y_0 \in B \mid x) - 1]$$
$$\leq P(y_m \in B \mid x) \leq \min[P(y_1 \in B \mid x), P(y_0 \in B \mid x)].$$

When the treatment policy is known to maximize $P(y_m \in B \mid x)$, we obtain the bound

$$(3.24) \quad \max[P(y_1 \in B \mid x), P(y_0 \in B \mid x)] \leq P(y_m \in B \mid x)$$
$$\leq \min[P(y_1 \in B \mid x) + P(y_0 \in B \mid x), 1].$$

It is interesting to compare these bounds with those under other assumptions. In (3.23), the lower bound coincides with the lower bound in the absence of prior information, while the upper bound coincides with the lower bound under the assumption that treatment is independent of the outcomes. In (3.24), the lower bound coincides with the upper bound under the assumption that treatment is independent of the outcomes, while the upper bound coincides with the upper bound in the absence of prior information. Thus the three treatment policies examined here imply that $P(\gamma_m \in B \mid x)$ lies in mutually exclusive intervals, and these three intervals partition the range of values that is feasible in the absence of prior information.

The idea of optimizing treatments has important applications in economics and in survival analysis. Economic analyses of voluntary treatment policies often assume that the treatment yielding the larger outcome is selected, so

$$(3.25) \quad \gamma_m = \max(\gamma_1, \gamma_0).$$

An example is the Roy model of occupation choice, discussed previously in Section 2.6. For any t, treatment policy (3.25) makes $P(\gamma_m \leq t \mid x) = P(\gamma_1 \leq t \cap \gamma_0 \leq t \mid x)$. So this policy minimizes $P(\gamma_m \leq t \mid x)$. We may therefore apply (3.23) to show that

$$(3.26) \quad \max[0, P(\gamma_1 \leq t \mid x) + P(\gamma_0 \leq t \mid x) - 1]$$
$$\leq P(\gamma_m \leq t \mid x) \leq \min[P(\gamma_1 \leq t \mid x), P(\gamma_0 \leq t \mid x)].$$

The competing-risks model of survival analysis assumes that the treatment yielding the smaller outcome is selected, so

$$(3.27) \quad \gamma_m = \min(\gamma_1, \gamma_0).$$

For any t, this treatment policy maximizes $P(\gamma_m \leq t \mid x)$. So (3.24) shows that

$$(3.28) \quad \max[P(\gamma_1 \leq t \mid x), P(\gamma_0 \leq t \mid x)] \leq P(\gamma_m \leq t \mid x)$$
$$\leq \min[P(\gamma_1 \leq t \mid x) + P(\gamma_0 \leq t \mid x), 1].$$

Known Treatment Distribution

The restrictions on treatment policy examined so far in this section specify the rule used to make treatment assignments, but do not constrain the fraction of the population receiving each treatment. It is also of interest to consider the reverse situation, where one knows the fraction receiving each treatment but does not know the rule used to make treatment assignments. For example, we noted earlier that resource constraints could limit implementation of the Perry Pre-school treatment to part of the eligible population. Knowledge of the budget constraint and the cost of preschool would suffice to determine the fraction of the population receiving the treatment. It may be more difficult to learn how school officials, social workers, and parents in-teract to determine which children receive the treatment.

Thus suppose that under policy m, a known fraction p of the persons with covariates x receive treatment 0 and the remaining fraction $1 - p$ receive treatment 1. So

$$(3.29) \quad P(z_m = 0 \mid x) = p,$$

where p is known. No information is available on the rule used to make treatment assignments that satisfy (3.29).

Given (3.29), $P(y_m \mid x)$ may be written

$$(3.30) \quad P(y_m \mid x) = P(y_1 \mid x, z_m = 1)(1 - p) + P(y_0 \mid x, z_m = 0)p.$$

The distributions $[P(y_1 \mid x), P(y_0 \mid x)]$ may be written

$$(3.31a) \quad P(y_1 \mid x) = P(y_1 \mid x, z_m = 1)(1 - p) + P(y_1 \mid x, z_m = 0)p$$

and

$$(3.31b) \quad P(y_0 \mid x) = P(y_0 \mid x, z_m = 1)(1 - p) + P(y_0 \mid x, z_m = 0)p.$$

Knowledge of $P(y_1 \mid x)$ and p restricts $P(y_1 \mid x, z_m = 1)$ and $P(y_1 \mid x, z_m = 0)$ to pairs of distributions that satisfy (3.31a); simi-larly, knowledge of $P(y_0 \mid x)$ and p restricts $P(y_0 \mid x, z_m = 1)$ and $P(y_0 \mid x, z_m = 0)$ to pairs of distributions that satisfy (3.31b). Through examination of the feasible pairs, it can be shown that $P(y_m \in B \mid x)$ satisfies the following sharp bound (see Manski, 1994b):

$$(3.32) \quad \max[0, P(y_1 \in B \mid x) - p] + \max[0, P(y_0 \in B \mid x) - (1 - p)]$$

$$\leq P(y_m \in B \mid x)$$

$$\leq \min[1 - p, P(y_1 \in B \mid x)]$$

$$+ \min[p, P(y_0 \in B \mid x)].$$

3.7. Identifying Combinations of Assumptions

Taken one at a time, the assumptions examined in Sections 3.5 and 3.6 improve the worst-case bound of Section 3.4, but are not strong enough to identify the outcome distribution under policy m. What assumptions do identify this distribution?

Suppose one combines the assumption that treatment is independent of outcomes with prior knowledge of the fraction of the population receiving each treatment. These two assumptions together imply that

$$(3.33) \quad P(y_m \mid x) = P(y_1 \mid x)(1 - p) + P(y_0 \mid x)p,$$

where p is the known fraction of the population receiving treatment 0. All the quantities on the right side are identified, so $P(y_m \mid x)$ is identified.

Alternatively, suppose that the outcomes y_1 and y_0 are statistically independent of one another and that the treatment with the larger outcome is always selected. Then

$$(3.34) \quad P(y_m \leq t \mid x) = P[\max(y_1, y_0) \leq t \mid x]$$

$$= P(y_1 \leq t \mid x) P(y_0 \leq t \mid x)$$

is identified for all values of t. Thus $P(y_m \mid x)$ is identified.

4

Response-Based Sampling

In the Introduction I quoted a standard text on epidemiology as stating that retrospective studies of disease are "useless from the point of view of public health," but "valid from the more general point of view of the advancement of knowledge" (Fleiss, 1981, p. 92). The term "retrospective studies" refers to a sampling process also known to epidemiologists as *case-control* sampling. This sampling process is known to economists studying individual behavior as *choice-based* sampling. I shall use the discipline-neutral term *response-based* sampling here.

Consider a population each of whose members is described by some covariates x and by a binary response y. A common task in empirical research in epidemiology, economics, and elsewhere is to infer the response probabilities $P(y \mid x)$ when the population is divided into response strata and random samples are drawn from one or both strata. This is response-based sampling.

Sampling from the stratum with $y = 1$ reveals the distribution $P(x \mid y = 1)$ of covariates within this stratum. Sampling from the stratum with $y = 0$ reveals $P(x \mid y = 0)$. Response-based sampling thus raises this identification question: What does knowledge of $P(x \mid y = 1)$ and/or $P(x \mid y = 0)$ reveal about $P(y \mid x)$?

This chapter examines the problem of inference from response-based samples and then uses the findings to study more general forms of stratified sampling. I begin by describing the epidemiological research practices that motivate Fleiss's curious statement.

4.1. The Odds Ratio and Public Health

Relative and Attributable Risk

Let $x = (w, r)$, where w denotes some covariates and r denotes other covariates referred to as a *risk factor* for a specified disease. A classical epidemiological problem is to learn how the value of the risk factor r affects the prevalence of the disease among persons with covariates w. Let $r = j$ and $r = k$ indicate any two values of the risk factor. Let the presence or absence of the disease be indicated by a binary variable y, with $y = 1$ if a person is ill and $y = 0$ if healthy. Let $P(y = 1 \mid w, r = k)$ be the probability of illness for a person with covariates w who has value k of the risk factor and let $P(y = 1 \mid w, r = j)$ be the probability of illness for a person with covariates w who has value j of the risk factor. The problem is to compare $P(y = 1 \mid w, r = k)$ and $P(y = 1 \mid w, r = j)$.

Epidemiologists compare these conditional disease probabilities through the *relative risk*

$$(4.1) \qquad RR \equiv \frac{P(y = 1 \mid w, r = k)}{P(y = 1 \mid w, r = j)}$$

and the *attributable risk*

$$(4.2) \qquad AR \equiv P(y = 1 \mid w, r = k) - P(y = 1 \mid w, r = j).$$

For example, let y indicate the occurrence of heart disease, let r indicate whether a person smokes cigarettes (yes $= k$, no $= j$), and let w give a person's age, sex, and occupation. For each value of w, RR gives the ratio of the probability of heart disease conditional on smoking to the probability of heart disease conditional on not smoking, while AR gives the difference between these probabilities.

Texts on epidemiology discuss both relative and attributable risk, but empirical research has focused on relative risk. This focus is hard to justify from the perspective of public health. The health impact of altering a risk factor presumably depends on the number of illnesses averted; that is, on the attributable risk times the size of the population. The relative risk statistic is uninformative about this quantity.

To illustrate, consider two scenarios. In one, the probability of heart disease conditional on smoking is .12 and conditional on non-smoking is .08. In the other, these probabilities are .00012 and .00008. The relative risk in both scenarios is 1.5. The attributable risk is .04 in the first scenario and .00004 in the second.

It seems odd that epidemiological research emphasizes relative risk rather than attributable risk. Indeed, the practice has long been criticized (see Berkson, 1958; Fleiss, 1981, section 6.3; and Hsieh, Manski, and McFadden, 1985). The rationale, such as it is, seems to rest on the widespread use in epidemiology of response-based sampling.

Random sampling of the population reveals the distribution $P(y, w, r)$, hence the response probabilities $P(y \mid w, r = k)$ and $P(y \mid w, r = j)$. Epidemiologists have found, however, that random sampling can be a costly way to gather data. Therefore they have often turned to less expensive stratified sampling designs, and especially to response-based designs. One divides the population into ill ($y = 1$) and healthy ($y = 0$) response strata and samples at random within each stratum. Response-based designs are considered to be particularly cost-effective in generating observations of serious diseases, because ill persons are clustered in hospitals and other treatment centers.

Cost-effectiveness in data collection is a virtue only if the sampling process reveals something useful to the researcher. Response-based sampling identifies the conditional distributions $P(w, r \mid y = 1)$ and $P(w, r \mid y = 0)$, which are not of direct interest to epidemiologists. How then have epidemiologists used response-based samples?

Bayes Theorem plays a large role in the analysis of response-based sampling, so it simplifies the exposition to assume that the covariates x have a discrete distribution. I shall maintain this assumption throughout the chapter, but the reader should understand that it is not essential. To apply the analysis to situations in which some components of x have continuous distributions, one need only replace probabilities of x values by densities in the statement of Bayes Theorem.

The Rare-Disease Assumption

Let us begin with an important negative fact. Consider any value of (w, r) with $P(w, r \mid y = 1) > 0$ and $P(w, r \mid y = 0) > 0$. Response-based sampling data alone reveal nothing about the magnitude of the response probability $P(y = 1 \mid w, r)$.

To see this, use Bayes Theorem to write

(4.3) $P(y = 1 \mid w, r)$

$$= \frac{P(w, r \mid y = 1) P(y = 1)}{P(w, r)}$$

$$= \frac{P(w, r \mid y = 1) P(y = 1)}{P(w, r \mid y = 1) P(y = 1) + P(w, r \mid y = 0) P(y = 0)}.$$

Response-based sampling identifies both $P(w, r \mid y = 1)$ and $P(w, r \mid y = 0)$, but provides no information about $P(y = 1)$. The fact that $P(y = 1)$ can lie anywhere between zero and one implies that $P(y = 1 \mid w, r)$ can lie anywhere between zero and one.[1]

Facing this situation, epidemiologists have commonly combined response-based sampling data with the assumption that the disease under study occurs rarely in the population. Formally, analysis under the *rare-disease assumption* is concerned with the limiting behavior of relative and attributable risk as $P(y = 1 \mid w)$ approaches zero.

The rare-disease assumption identifies both relative and attributable risk. To see this, rewrite equation (4.3) in the equivalent form

(4.4) $P(y = 1 \mid w, r)$

$$= \frac{P(r \mid w, y = 1) P(y = 1 \mid w)}{P(r \mid w)}$$

$$= \frac{P(r \mid w, y = 1) P(y = 1 \mid w)}{P(r \mid w, y = 1) P(y = 1 \mid w) + P(r \mid w, y = 0) P(y = 0 \mid w)}.$$

Inserting the right-side expression into the definitions of relative and attributable risk yields

(4.5)

$$RR =$$

$$\left[\frac{P(r = k \mid w, y = 1)}{P(r = j \mid w, y = 1)}\right.$$

$$\left.\times \frac{P(r = j \mid w, y = 1)\,P(y = 1 \mid w) + P(r = j \mid w, y = 0)\,P(y = 0 \mid w)}{P(r = k \mid w, y = 1)\,P(y = 1 \mid w) + P(r = k \mid w, y = 0)\,P(y = 0 \mid w)}\right]$$

and

(4.6)

$$AR =$$

$$\left[\frac{P(r = k \mid w, y = 1)\,P(y = 1 \mid w)}{P(r = k \mid w, y = 1)\,P(y = 1 \mid w) + P(r = k \mid w, y = 0)\,P(y = 0 \mid w)}\right.$$

$$\left.- \frac{P(r = j \mid w, y = 1)\,P(y = 1 \mid w)}{P(r = j \mid w, y = 1)\,P(y = 1 \mid w) + P(r = j \mid w, y = 0)\,P(y = 0 \mid w)}\right].$$

Letting $P(y = 1 \mid w)$ approach zero, these expressions reduce to

$$(4.7) \qquad RR = \frac{P(r = k \mid w, y = 1)}{P(r = j \mid w, y = 1)} \; \frac{P(r = j \mid w, y = 0)}{P(r = k \mid w, y = 0)}$$

and

$$(4.8) \qquad AR = 0.$$

Cornfield (1951) showed that equation (4.7) is the relative risk under the rare-disease assumption. The expression on the right side of (4.7) is called the *odds ratio* and may also be written as a function of the response probabilities. That is,

$$(4.9) \qquad OR \equiv \frac{P(y = 1 \mid w, r = k)}{P(y = 0 \mid w, r = k)} \; \frac{P(y = 0 \mid w, r = j)}{P(y = 1 \mid w, r = j)}$$

$$= \frac{P(r = k \mid w, y = 1)\ P(r = j \mid w, y = 0)}{P(r = j \mid w, y = 1)\ P(r = k \mid w, y = 0)}.$$

Equality of these expressions follows from equation (4.4).

Cornfield's finding motivates the widespread epidemiological practice of using response-based samples to estimate the odds ratio and then invoking the rare-disease assumption to interpret the odds ratio as relative risk. Fleiss's statement that retrospective studies are "valid from the more general point of view of the advancement of knowledge" is an endorsement of this practice, despite its associated implication that attributable risk is zero. Fleiss's statement that retrospective studies are "useless from the point of view of public health" reflects the widespread belief that (Fleiss, 1981, p. 92) "retrospective studies are incapable of providing estimates" of attributable risk. I show in the next section that this assessment is somewhat too pessimistic.

4.2. Bounds on Relative and Attributable Risk

Suppose that a pair of response-based samples is available, with no other information. Although response-based sampling reveals nothing about the magnitude of the response probability $P(y = 1 \mid w, r)$ at a fixed value of (w, r), this sampling process is informative about the way that $P(y = 1 \mid w, r)$ varies with r. We already know that the odds ratio is identified. Inspection of equation (4.9) shows that the odds ratio reveals whether $P(y = 1 \mid w, r = k)$ is larger than $P(y \mid w, r = j)$. In particular,

(4.10a) OR $< 1 \Rightarrow P(y = 1 \mid w, r = k) < P(y = 1 \mid w, r = j)$

(4.10b) OR $= 1 \Rightarrow P(y = 1 \mid w, r = k) = P(y = 1 \mid w, r = j)$

(4.10c) OR $> 1 \Rightarrow P(y = 1 \mid w, r = k) > P(y = 1 \mid w, r = j)$.

We can go beyond (4.10) to prove that response-based sampling implies informative lower and upper bounds on relative and attributable risks. These bounds are developed here.

I use a numerical example concerning smoking and heart disease to illustrate the findings. Among persons with specified covariates w, let the actual probabilities of heart disease conditional on smoking and nonsmoking be .12 and .08, and let the fraction of persons who smoke be .50. These values imply that the unconditional probability of heart disease is .10 and that the probabilities of smoking conditional on being ill and healthy are .60 and .49. The implied odds ratio is 1.57, relative risk is 1.50, and attributable risk is .04. Thus the parameters of the example are

$$P(y = 1 \mid w, r = k) = .12 \qquad P(y = 1 \mid w, r = j) = .08$$

$$P(y = 1 \mid w) = .10 \qquad\qquad P(r = k \mid w, y = 1) = .60$$

$$OR = 1.57 \qquad\qquad\qquad RR = 1.50$$

$$P(r = k \mid w) = P(r = j \mid w) = .50$$

$$P(r = k \mid w, y = 0) = .49$$

$$AR = .04.$$

Bound on Relative Risk

Examine the expression for relative risk given in equation (4.5). All the quantities on the right side of this equation are identified by response-based sampling except for $P(y \mid w)$. All that is known is that $P(y = 1 \mid w)$ and $P(y = 0 \mid w)$ are nonnegative and sum to one. So we may determine the range of possible values for RR by analyzing how the right side of (4.5) varies across the logically possible values of $P(y \mid w)$. The result is that the relative risk must lie between the odds ratio and the value 1. That is, when $P(w, r \mid y = 1)$ and $P(w, r \mid y = 0)$ are known, sharp bounds on RR are[2]

(4.11a) $OR < 1 \Rightarrow OR \leq RR \leq 1$

(4.11b) $OR = 1 \Rightarrow RR = 1$

(4.11c) $OR > 1 \Rightarrow 1 \leq RR \leq OR.$

In our example concerning smoking and heart disease, the odds ratio is 1.57. So we may conclude that the probability of heart disease conditional on smoking is at least as large as but no more than 1.57 times the probability conditional on nonsmoking. Recall that the rare-disease assumption makes relative risk equal the odds ratio. Thus this conventional epidemiological assumption always makes relative risks appear further from one than they actually are. The magnitude of the bias depends on the actual prevalence of the disease under study. In particular, the bias grows as $P(y = 1 \mid w)$ moves away from zero.[3] In the smoking–heart disease example, the bias is small because the actual relative risk is 1.50.

Bound on Attributable Risk

Examine the expression for attributable risk given in equation (4.6). Again, all the quantities on the right side are identified by response-based sampling except for $P(y \mid w)$. So we may determine the range of possible values for AR by analyzing how the right side of (4.6) varies across the logically possible values of $P(y \mid w)$.

Let AR_p denote the value that the attributable risk would take if $P(y = 1 \mid w)$ were to equal any value p. That is, define

$$(4.12) \quad AR_p \equiv$$

$$\frac{P(r = k \mid w, y = 1)p}{P(r = k \mid w, y = 1)p + P(r = k \mid w, y = 0)(1 - p)}$$

$$- \frac{P(r = j \mid w, y = 1)p}{P(r = j \mid w, y = 1)p + P(r = j \mid w, y = 0)(1 - p)}.$$

The result is that AR must lie between AR_π and zero, where

$$(4.13) \quad \pi \equiv$$

$$\frac{\beta P(r{=}k \mid w, y{=}0) - P(r{=}j \mid w, y{=}0)}{[\beta P(r{=}k \mid w, y{=}0) - P(r{=}j \mid w, y{=}0)] - [\beta P(r{=}k \mid w, y{=}1) - P(r{=}j \mid w, y{=}1)]}$$

and where

(4.14)
$$\beta \equiv \left[\frac{P(r = j \mid w, \, y = 1) \, P(r = j \mid w, \, y = 0)}{P(r = k \mid w, \, y = 1) \, P(r = k \mid w, \, y = 0)} \right]^{1/2}.$$

That is, when $P(w, r \mid y = 1)$ and $P(w, r \mid y = 0)$ are known, sharp bounds on AR are[4]

(4.15a) $\mathrm{OR} < 1 \Rightarrow \mathrm{AR}_\pi \leq \mathrm{AR} \leq 0$

(4.15b) $\mathrm{OR} = 1 \Rightarrow \mathrm{AR} = 0$

(4.15c) $\mathrm{OR} > 1 \Rightarrow 0 \leq \mathrm{AR} \leq \mathrm{AR}_\pi.$

In our heart disease example, $\beta = .83$, $\pi = .51$, and $\mathrm{AR}_\pi = .11$. Hence response-based sampling reveals that the attributable risk associated with smoking is between 0 and .11. (Recall that .04 is the actual value, which is unknown to the researcher.) This seems a useful finding from a public health perspective. After all, in the absence of empirical evidence, AR could take any value between -1 and 1.

4.3. Information on Marginal Distributions

The problem of inference from response-based samples is that this sampling process does not reveal the marginal response distribution $P(y)$. If $P(y)$ were known, then we would have all the information needed to identify the joint distribution $P(y, w, r)$; hence the response probabilities $P(y \mid w, r)$ would be identified.

Beginning with Manski and Lerman (1977), the econometric literature on response-based sampling has emphasized that it is often possible to learn $P(y)$ from auxiliary data sources. Hsieh, Manski, and McFadden (1985) point out that published health statistics drawn from national household surveys or from hospital administrative records provide estimates of the population prevalence of many diseases. This article also calls attention to the fact that $P(y)$ may be inferred from information on the marginal distribution of the risk factor or of the other covariates.

The marginal probability that the risk factor takes any value i is

$$(4.16) \quad P(r = i) = P(r = i \mid y = 1) P(y = 1)$$
$$+ P(r = i \mid y = 0) [1 - P(y = 1)].$$

Response-based sampling identifies both $P(r = i \mid y = 1)$ and $P(r = i \mid y = 0)$. Suppose that $P(r = i)$ is known. Then (4.16) implies that $P(y)$ is identified, provided only that $P(r = i \mid y = 1) \neq P(r = i \mid y = 0)$. For example, in a study of smoking and heart disease, national survey data may reveal the fraction of the population who are smokers. This information may then be combined with response-based sampling data to learn the fraction of the population with heart disease.

Similar reasoning shows that $P(y)$ may be inferred from knowledge of the expected value of some covariate. For example, let the first component of w be a person's age. Suppose that the average age $E(w_1)$ of the population is known. Observe that

$$(4.17) \quad E(w_1) = E(w_1 \mid y = 1) P(y = 1)$$
$$+ E(w_1 \mid y = 0) [1 - P(y = 1)].$$

Response-based sampling identifies $E(w_1 \mid y = 1)$ and $E(w_1 \mid y = 0)$. Hence knowledge of $E(w_1)$ implies knowledge of $P(y)$, provided only that the average age $E(w_1 \mid y = 1)$ of ill people is not the same as the average age $E(w_1 \mid y = 0)$ of healthy ones.

4.4. Sampling from One Response Stratum

The literature on response-based sampling has concentrated on situations in which one samples at random from both response strata, and so learns both $P(w, r \mid y = 1)$ and $P(w, r \mid y = 0)$. Often, however, one is only able to sample from a single response stratum, say, from the subpopulation with $y = 1$, and so learns only $P(w, r \mid y = 1)$. For example, an epidemiologist studying the prevalence of a disease may use hospital records to learn the distribution of covariates among persons who are ill ($y = 1$), but may have no comparable data on persons who are healthy ($y = 0$). A social policy analyst studying participation in welfare programs may use the administrative records of the welfare system to learn the backgrounds of welfare recipients ($y = 1$), but may have no comparable information on nonrecipients ($y = 0$).

Sampling from one response stratum obviously reveals nothing about the magnitude of response probabilities. Nor does it reveal anything about relative and attributable risks. But inference becomes possible if auxiliary distributional information is available.

Recall equations (4.3) and (4.4), where Bayes Theorem was used to write the response probabilities in the equivalent forms

$$P(y = 1 \mid w, r) = \frac{P(w, r \mid y = 1) P(y = 1)}{P(w, r)}$$

$$= \frac{P(r \mid w, y = 1) P(y = 1 \mid w)}{P(r \mid w)}.$$

Sampling from the stratum with $y = 1$ reveals $P(w, r \mid y = 1)$. Hsieh, Manski, and McFadden (1985) point out that the response probability $P(y = 1 \mid w, r)$ is identified if auxiliary data sources reveal the marginal response distribution $P(y)$ and the distribution $P(w, r)$ of covariates. This result also holds if auxiliary data sources reveal the conditional distributions $P(y \mid w)$ and $P(r \mid w)$. An empirical illustration follows.

Using Administrative Records to Infer AFDC Transition Rates

Public concern with the perceived problem of "welfare dependence" in the United States has stimulated considerable recent research on the dynamics of participation in the federal program Aid for Families with Dependent Children. Much of this work is described in the *Green Book,* an annual volume overviewing government entitlement programs prepared by the Committee on Ways and Means of the U.S. House of Representatives (U.S. House of Representatives, 1993).

Empirical study of the dynamics of AFDC participation is conceptually straightforward if one can draw a random sample of the United States population and follow the respondents over their lives. Then

one can, in principle, learn the complete time path of welfare participation of each respondent. Two surveys that approach this ideal are
the Panel Study of Income Dynamics (PSID), begun in 1968 by the
Institute for Social Research at the University of Michigan, and the
National Longitudinal Survey of Youth (NLSY), begun in 1979 by
the U.S. Department of Labor. Research on the dynamics of AFDC
has been based primarily on PSID and NLSY data.

The high cost of administering longitudinal surveys such as the
PSID and NLSY inhibits the drawing of samples large enough to reach
precise statistical conclusions about welfare dynamics. It therefore
makes sense to ask what may be learned from less costly response-
based designs. Moffitt (1992b) observes that the administrative records
of the welfare system make it inexpensive to sample from the stratum
of active AFDC recipients, but that it is much more costly to sample
from the stratum of nonrecipients. He suggests studying the dynamics
of AFDC participation by sampling from the former response stratum
alone.

I shall formalize this suggestion and then apply the simple result
pointed out by Hsieh, Manski, and McFadden (1985). Let T_0 be a
specified date in time and let T_1 be a specified later date. Let w denote
specified covariates. Let

$r = k$ if a family receives AFDC payments at date T_0;
$\quad = j$ otherwise.
$y = 1$ if a family receives AFDC payments at date T_1;
$\quad = 0$ otherwise.

The dynamics of AFDC participation between T_0 and T_1 are ex-
pressed by the *transition probabilities* $P(y = 1 \mid w, r = k)$ and $P(y = 1 \mid w, r = j)$.

Data on AFDC recipiency collected by the Administration for
Children and Families of the U.S. Department of Health and Human
Services and reported in the *1993 Green Book* provide the basis for
"back of the envelope" estimates of $P(r \mid w, y = 1)$, $P(y \mid w)$ and
$P(r \mid w)$. Among the families receiving AFDC in some month of the
year 1991, 64.7 percent had continuously received payments during
the previous twelve months (*1993 Green Book*, section 7, table 35,
p. 705). Let T_1 be a date in 1991 and let T_0 be the same date twelve

months earlier. Let the covariate w indicate a family with children under age 18 in the year 1991. Assume that all of the families receiving AFDC in 1990 or 1991 had children under age 18 in 1991. Also assume that any family receiving AFDC at both of the dates T_0 and T_1 received payments during the entire twelve-month interim period. Then the *Green Book* data reveal that $P(r = k \mid w, y = 1) = .647$.

The average monthly number of families receiving AFDC was 3,974,000 in 1990 and 4,375,000 in 1991 (*1993 Green Book,* section 7, table 1, p. 616). There were 34,973,000 families with children under age 18 in 1991 (*1993 Green Book,* appendix G, table 4, p. 1117). So the *Green Book* data reveal that $P(r = k \mid w) = 3,974,000/ 34,973,000 = .114$ and $P(y = 1 \mid w) = 4,375,000/34,973,000 = .125$.

Combining these findings yields

$$P(y = 1 \mid w, r = k) = \frac{P(r = k \mid w, y = 1)\,P(y = 1 \mid w)}{P(r = k \mid w)} = .709$$

$$P(y = 1 \mid w, r = j) = \frac{P(r = j \mid w, y = 1)\,P(y = 1 \mid w)}{P(r = j \mid w)} = .050.$$

Thus a family receiving AFDC in 1990 had a .709 chance of receiving AFDC in 1991, but a family not receiving AFDC in 1990 had only a .050 chance of receiving AFDC in 1991.

4.5. General Binary Stratifications

Analysis of response-based sampling provides the foundation for study of more general stratified sampling processes in which a population is divided into two strata and random samples are drawn from one or both strata. The strata need not coincide with values of the outcome y. Nor need y be a binary variable.

To describe general binary stratifications, let each member of the population be described by values for (y, x, s). As before, (y, x) are the outcome and covariates. The new variable s indicates the stratum containing each person, with $s = 1$ if a person is a member of stratum 1 and $s = 0$ otherwise. Random sampling from stratum 1 identifies the distribution $P(y, x \mid s = 1)$, while random sampling from stratum 0 identifies $P(y, x \mid s = 0)$.

Our concern is to infer the conditional distribution $P(y \mid x)$. The analysis of relative and attributable risk in earlier sections of this chapter may be extended from response-based sampling to general binary stratified sampling. I shall examine only the problem of inference on $P(y \mid x)$ at a specified value of x, however. Hence we no longer need decompose x into the covariates w and the risk factor r.

Sampling from Both Strata

Let us first examine the situation in which samples are drawn from both strata. Write $P(y \mid x)$ as

$$(4.18) \quad P(y \mid x) = P(y \mid x, s = 1) P(s = 1 \mid x)$$
$$+ P(y \mid x, s = 0) P(s = 0 \mid x).$$

The sampling process identifies $P(y \mid x, s = 1)$ and $P(y \mid x, s = 0)$. The sampling process does not restrict the stratum distribution $P(s \mid x)$, because inference on $P(s \mid x)$ is precisely the response-based sampling problem, with s here in the role of y earlier. So sampling from both strata reveals $P(y \mid x)$ to be an unrestricted mixture of the two distributions $P(y \mid x, s = 1)$ and $P(y \mid x, s = 0)$.

How informative is the sampling process about $P(y \mid x)$? The answer depends on how closely related the outcomes y and strata s are to one another, conditional on x. At one extreme, suppose that y and s are statistically independent conditional on x; that is, $P(y \mid x, s = 1) = P(y \mid x, s = 0)$. Then equation (4.18) reduces to

$$(4.19) \quad P(y \mid x) = P(y \mid x, s = 1) = P(y \mid x, s = 0),$$

whatever $P(s \mid x)$ may be. Thus $P(y \mid x)$ is identified. Stratification is said to be *exogenous* when (4.19) holds.

At the other extreme, suppose that y is a binary response and that the sampling process is response-based. Then $P(y = 1 \mid x, s = 1) = 1$, $P(y = 1 \mid x, s = 0) = 0$, and equation (4.18) reduces to

$$(4.20) \quad P(y \mid x) = P(s \mid x).$$

The sampling process does not restrict $P(y \mid x)$ at all. We may conclude that, from the perspective of inference on $P(y \mid x)$ at a specified value of x, response-based sampling is the least informative member of the class of binary stratified sampling processes.

This discussion has presumed that no information other than the stratified samples is available. The analysis of Section 4.3 shows that $P(s \mid x)$ is identified if auxiliary data reveal either the marginal stratum distribution $P(s)$ or the distribution $P(x)$ of covariates. Combining stratified samples with such auxiliary data identifies $P(y \mid x)$.

Sampling from One Stratum

Now suppose that observations are drawn from only one stratum, say, from the stratum where $s = 1$. Then the sampling process no longer identifies $P(y, x \mid s = 0)$. Inspection of equation (4.18) shows that there are no implied restrictions on $P(y \mid x)$. As $P(s = 1 \mid x)$ approaches zero, $P(y \mid x)$ approaches the unrestricted $P(y \mid x, s = 0)$.

Researchers sampling from one stratum commonly identify $P(y \mid x)$ by assuming that stratification is exogenous. An alternative is to bring to bear auxiliary distributional information that, in the manner of Section 4.4, reveals the stratum distribution $P(s \mid x)$. Suppose that auxiliary information reveals $P(s \mid x)$. Then $P(y \mid x, s = 1)$ and $P(s \mid x)$ are identified, but $P(y \mid x, s = 0)$ is unrestricted.

This is the same information that is available in the selection problem examined in Chapter 2. Here $P(y \mid x, s = 1)$ is the distribution of outcomes conditional on selection, $P(s = 1 \mid x)$ is the selection probability, and $P(y \mid x, s = 0)$ is the distribution of outcomes conditional on censoring. So the problem of inference when a random sample drawn from one stratum is combined with knowledge of $P(s \mid x)$ is equivalent to the problem of inference under censored sampling.

5

Predicting Individual Behavior

5.1. Revealed Preference Analysis

Suppose that a person must choose among three alternatives (A, B, C) and is observed to choose A. Now let this person be faced with a new decision problem, in which he or she must choose between A and B, alternative C not being available. Can the decision be predicted?

This is an extrapolation problem, so the answer must depend on what one is willing to assume. A mainstream economist would predict that the chosen alternative remains A. The reasoning is that the observed selection of A over B and C reveals this person to prefer A to B. Removing alternative C from the choice set does not disturb this fact.

This is perhaps the simplest application of *revealed preference analysis,* the approach economists routinely use to predict individual behavior. To make predictions, economists combine empirical evidence with behavioral assumptions. The data are observations of actual choices made by members of the population of interest. The basic behavioral assumption is *rational choice:* each member of the population orders alternatives in terms of preference and chooses the most preferred alternative among those available.

The assumption of rational choice suffices to make some extrapolations from observed choices. One has already been given. Another concerns a person whose behavior is observed in two decision settings. In the first setting, the person must choose between A and B, and is observed to choose A. In the second setting, the same person must choose between B and C, and is observed to choose B. *Transitivity* of preferences implies that if this person were required to choose between A and C, he or she would choose A.

More ambitious extrapolations require stronger assumptions than rational choice alone. Economists often observe the choices made by a sample of persons in certain decision settings, and wish to predict the choices that would be made by other members of the population in other decision settings. Extrapolations of this kind become possible when rational choice is accompanied by assumptions restricting the form of preferences.

This chapter begins by describing the practice of revealed preference analysis. Revealed preference analysis needs to be well understood even by those who do not accept its premises. Other social scientists often criticize as implausible the behavioral assumptions underlying econometric predictions. But criticism in the absence of a constructive alternative has only limited effectiveness. The other social sciences have not developed any coherent and widely applicable competitor to revealed preference analysis, and thus this approach continues to dominate efforts to extrapolate individual behavior.

College Choice in America

I shall use my own revealed preference analysis of college-going behavior, performed with David Wise, to describe the main features of present-day econometric practice. Manski and Wise (1983, chaps. 6 and 7) used observations from the National Longitudinal Study of the High School Class of 1972 (NLS72) to estimate a rational-choice model of college enrollment. We then used the estimated model to predict the impact of the Pell Grant program, the major federal college scholarship program, on enrollment.

The NLS72 survey, commissioned by the National Center for Education Statistics, provides schooling, work, and background data on almost 23,000 high school seniors drawn from over 1,300 high schools in the United States in spring 1972. Data were obtained through a series of questionnaires distributed to the respondents and their high schools and through periodic follow-up surveys.

The starting point for our analysis was to assume that the patterns of college enrollment and labor force participation observed among the NLS72 respondents are the consequence of decisions made by these students, by colleges, and by employers. Colleges and employers make admissions decisions and job offers that determine the options available to each high school senior upon graduation. Each senior selects among the available options.

What do the NLS72 data reveal about the decision processes generating postsecondary activities? If we assume that a student chooses the most preferred alternative from the available options, then observations of chosen activities partially reveal student preferences. In particular, if we imagine a student as implicitly assigning a numerical utility value to each potential activity, then the fact that the student has chosen a particular activity implies that its utility exceeds that of all others that the student could have chosen.

For simplicity, assume that after high school graduation a student has two alternatives: $y = 1$ is college enrollment and $y = 0$ is work. (The model actually estimated assumed multiple alternatives.) If we observe that NLS72 respondent n chose to go to college, then we may infer that $U_{n1} \geq U_{n0}$, where U_{n1} and U_{n0} are the utilities that this respondent associated with college and work. If respondent n chose to work, then $U_{n0} \geq U_{n1}$. The NLS72 data provide a large set of these inequalities, one for each respondent.

The preference inequalities implied by observations of these activity choices do not provide sufficient information to allow us to predict how a student not in the sample would select between college and work, or how a student in the sample would have behaved if conditions had differed. To extrapolate behavior, we must combine the NLS72 sample information with assumptions restricting the form of preferences.

For example, we might assume that the utility of college enrollment to student n depends on his ability A_n and his parents' income I_n, and also on the quality Q_n and cost C_n of his best college opportunity. Similarly, the utility of working might depend on his best potential wage W_n. In particular, suppose that the utilities are known to have the linear form

(5.1a) $U_{n1} = \beta_1 A_n + \beta_2 I_n + \beta_3 Q_n + \beta_4 C_n$

and

(5.1b) $U_{n0} = \beta_5 W_n,$

where $\beta \equiv \beta_1, \ldots, \beta_5$ is a parameter vector that is unknown but assumed not to equal zero.[1] Observe that β is assumed not to vary across students. For now, all students with the same value of (A, I, Q, C, W) are assumed to have the same preferences.

Given data on the actual activity y_n chosen by student n and data measuring the utility determinants $(A_n, I_n, Q_n, C_n, W_n)$, the revealed preference inequality

(5.2) $y_n = 1 \Rightarrow \beta_1 A_n + \beta_2 I_n + \beta_3 Q_n + \beta_4 C_n \geq \beta_5 W_n$

$\quad\quad\quad y_n = 0 \Rightarrow \beta_1 A_n + \beta_2 I_n + \beta_3 Q_n + \beta_4 C_n \leq \beta_5 W_n$

implies restrictions on the possible values of the parameters β. The NLS72 data yield an inequality of the form (5.2) for each respondent, and β must satisfy all of these inequalities.

Suppose that only one nonzero value of β satisfies all of the revealed preference inequalities implied by the NLS72 choice data. Then this must be the actual value of β. We can now predict the schooling/work decisions of students whose values of (A, I, Q, C, W) differ from those actually observed among the NLS72 respondents. For example, if i designates a student whose best college has attributes (A_i, I_i, Q_i, C_i) and whose best job opportunity has wage W_i, this student i may choose to work if

(5.3a) $\beta_5 W_i = \beta_1 A_i + \beta_2 I_i + \beta_3 Q_i + \beta_4 C_i$

and definitely chooses to work if

(5.3b) $\beta_5 W_i > \beta_1 A_i + \beta_2 I_i + \beta_3 Q_i + \beta_4 C_i.$

Random Utility Models and Choice Probabilities

The utility function of our example has a fairly simple form. An actual empirical study might pose determinants of utility beyond (A, I, Q, C, W) and loosen the assumption that these determinants act on utility in linear fashion; see, for example, the specification of Manski and Wise (1983). It is not reasonable to expect, however, that any empirical study will have access to data that express all the determinants of student decision making. Therefore economists assume that utilities vary as a function of variables that are known to decision makers but unobserved by the researcher.

In our example, these unobserved variables may be represented through real numbers u_{n1} and u_{n0} added to the utilities specified in (5.1). Thus we now generalize (5.1) to

$$\text{(5.4a)}\quad U_{n1} = \beta_1 A_n + \beta_2 I_n + \beta_3 Q_n + \beta_4 C_n + u_{n1}$$

and

$$\text{(5.4b)}\quad U_{n0} = \beta_5 W_n + u_{n0}.$$

With this change, the preference inequalities revealed by the NLS72 data become

$$\text{(5.5)}\quad y_n = 1 \Rightarrow \beta_1 A_n + \beta_2 I_n + \beta_3 Q_n + \beta_4 C_n - \beta_5 W_n + u_n \geq 0$$
$$y_n = 0 \Rightarrow \beta_1 A_n + \beta_2 I_n + \beta_3 Q_n + \beta_4 C_n - \beta_5 W_n + u_n \leq 0,$$

where $u_n \equiv u_{n1} - u_{n0}$.

From the perspective of the researcher, the unobserved u is a random variable with some distribution across the population of students. So the utilities U_{n1} and U_{n0} are themselves random variables, and we have a *random utility model* of behavior. Assuming that the distribution of u is continuous, the probability that a student with observed characteristics (A, I, Q, C, W) chooses to enroll in college is[2]

$$\text{(5.6)}\quad P(y = 1 \mid x) = P(x\beta + u \geq 0 \mid x),$$

where $x \equiv (A, I, Q, C, W)$. Equation (5.6) does not assert that a student with characteristics x actually behaves probabilistically. The equation asserts that a researcher is only able to make probabilistic predictions of student behavior.[3]

Given a random sample of realizations of (y, x), the choice probability $P(y = 1 \mid x)$ can be estimated nonparametrically on the support of $P(x)$. Our objective in deriving the random utility model representation of $P(y = 1 \mid x)$ on the right side of (5.6) is to enable prediction of y off the support.

A random utility model has no identifying power in the absence of information restricting the distribution of the unobserved variable u. Perhaps the easiest way to see this is through inspection of the revealed preference inequalities (5.5). Given any hypothesized value for the parameters β and any NLS72 respondent n, one can always find a value of u_n that satisfies (5.5). So the NLS72 choice data imply no restrictions on β.

The model becomes informative when sufficiently strong assumptions are imposed on the distribution of u across the population of students. The prevailing practice in empirical studies has been to assume that u is statistically independent of x, with a distribution known up to normalizations of location and scale.[4] In particular, suppose it is known that u has the standard logistic distribution. Then the choice probabilities have the form of the familiar *logit model*

$$(5.7) \quad P(y = 1 \mid x) = \frac{e^{x\beta}}{1 + e^{x\beta}}.$$

With the distribution of u specified in this way, the NLS72 data can be used to estimate the parameters β. This done, the right side of (5.7) can be used to predict the schooling behavior that would occur under conditions different from those faced by the NLS72 respondents.[5]

Predicting the Enrollment Effects of Student Aid Policy

Manski and Wise (1983) estimated a random utility model that is more complex than the one just described, but not qualitatively different. The estimated model was used to study the impact on freshman college enrollments of the Basic Educational Opportunity Grant (BEOG) program, later renamed the Pell Grant program. This federal scholarship program was initiated in 1973, so the NLS72 respondents were not eligible at the time of their initial postsecondary schooling decisions. Thus our prediction problem was clearly one of extrapolation rather than prediction on the support.

In the context of our random utility model, the BEOG program influences behavior by changing the college costs, C, that students face. Given knowledge of the program eligibility criteria and award formula, we estimated the cost of college to any given student in the presence of the program. This done, we predicted how the NLS72 respondents would have behaved had the program been in operation. We then aggregated these predictions across the sample to generate predictions of aggregate freshman college enrollments in the United States.

Table 5.1 presents some of our findings concerning the version of the program that was in effect in 1979. The predictions indicate that the BEOG program was responsible for a truly substantial increase (59 percent) in the college enrollment rate of low-income students, a moderate increase (12 percent) in middle-income enrollments, and a minor increase (3 percent) in the rate for upper-income students.

Overall, we predicted that 1,603,000 of the 3,300,000 persons who were high school seniors in 1979 would enroll in full-time post-secondary education in 1979–80. In contrast, only 1,324,000 would have enrolled had the BEOG program not been in operation. The findings indicate that the enrollment increases induced by the existence of the program were totally concentrated at two-year and vocational schools. Enrollments at four-year schools were essentially unaffected.

Table 5.1   Predicted enrollments in 1979 with and without the BEOG program (in thousands)

	Predicted enrollments by postsecondary school type							
	All schools		4-year		2-year		Voc-tech.	
Income group	With BEOG	Without BEOG	With BEOG	Without BEOG	With BEOG	Without BEOG	With BEOG	Without BEOG
Lower (below $16,900)	590	370	128	137	349	210	113	23
Middle	398	354	162	164	202	168	34	22
Upper (above $21,700)	615	600	377	378	210	198	28	24
Total	1603	1324	668	679	761	576	174	69

Power and Price of the Analysis

Federal scholarship programs with varying eligibility criteria and award formulae have been proposed, but only a few programs have actually been implemented. Our revealed preference analysis of college enrollments makes it possible to predict the impacts of a wide variety of proposed and actual programs. This ability to extrapolate is very powerful.

The price of extrapolation is the set of assumptions imposed. The assumption of rational choice alone yields only limited ability to extrapolate. The real power of revealed preference analysis emerges when the rational-choice assumption is combined with restrictions on the form of preferences. Our analysis of college choice indicates the kinds of preference assumptions commonly imposed in empirical studies.

5.2. How Do Youth Infer the Returns to Schooling?

I would like to call particular attention to one of the assumptions routinely imposed in applications of revealed preference analysis. Rational choice is a subjective concept. Individuals are assumed to choose the most preferred alternative, given their perceptions of the options available. There is no requirement that these perceptions be "correct" or "objective" in any sense.

Our discussion of revealed preference analysis has thus far avoided this critical matter by implicitly assuming that researchers know how decision makers perceive their alternatives. We assumed that high school graduates perceive their options to be college enrollment and labor force participation. We supposed that our measurements of student ability A, family income I, college quality Q and cost C, and potential wage W correspond to the way students perceive these factors.

There is often reason to question whether researchers really know how decision makers perceive their alternatives. I continue to focus on the case of schooling choice, which provides a good illustration.[6]

Prevailing Empirical Practices

Economists analyzing schooling decisions assume that youth compare the expected outcomes from schooling and other activities and then choose the best available option. Viewing education as an investment in human capital, we use the term *returns to schooling* to refer to the outcomes from schooling relative to nonschooling.

Given the centrality of the expected returns to schooling in economic thinking on schooling behavior, it might be anticipated that economists would make substantial efforts to learn how youth perceive the returns to schooling. But the profession has traditionally been skeptical of subjective data, so much so that we have generally been unwilling to collect data on expectations. Instead the norm has been to assume that expectations are formed in specific ways.

Economic studies of schooling behavior have routinely assumed that all youth possess the same information and process this information in the same manner. But researchers have differed substantially in their assumptions about the information that youth possess and the way they process it. I give three examples.

Freeman (1971) analyzed the major field decisions of male college students. He assumed that each person chooses the field offering the highest expected lifetime income. Moreover, he assumed that expectations formation is myopic. Each person believes that, should he select a given college major, he would obtain the mean income realized by the members of a specified earlier cohort who did make that choice.[7]

Willis and Rosen (1979) analyzed the college enrollment decision of male veterans of World War II. They assumed that a person chooses to enroll if the expected lifetime income associated with enrollment exceeds that of not enrolling. They assumed that youth condition their income expectations on their ability. They also assumed that youth have *rational expectations,* that is, that youth know the actual process generating lifetime incomes conditional on their ability and schooling. Willis and Rosen hypothesized that each youth applies his knowledge of the process generating lifetime incomes to predict his own future income should he enroll or not enroll in college.

In the Manski and Wise (1983, chap. 6) analysis of college choice, the utility of enrolling in a given college was assumed to depend on a student's own Scholastic Aptitude Test (SAT) score and on the average score of students enrolled in that college. Youth were not assumed to know the lifetime incomes realized by earlier cohorts or the actual process generating incomes. Instead they were assumed to believe that the returns to enrolling depend on their own SAT score and the average at the college.

Adolescent Econometricians

Unfortunately, there is no evidence supporting any of the assumptions economists have made about expectations of the returns to schooling. If anything, there are logical and empirical reasons to think otherwise.

The logical point is that youth seeking to infer their own returns to schooling face the same selection problem as do social scientists seeking to infer treatment effects. Youth and social scientists may possess different data on realized incomes, may have different knowledge of the economy, and may process their information in different ways. But both want to use their data and knowledge to learn the returns to schooling conditional on the available information. Youth, like social scientists, must confront the selection problem.

There is no empirical evidence that youth cope with the selection problem in any of the ways assumed in the studies cited. In fact, there is indirect evidence suggesting that youth form expectations in heterogeneous ways. Although we lack data on youths' expectations, we have extensive data on the practices of labor economists studying the returns to schooling. During the past thirty years, labor economists have executed hundreds of empirical studies of the returns to schooling. Reading this vast literature reveals that researchers vary greatly in the conditioning variables used, in the outcome data analyzed, and in the handling of the selection problem.

Compare, for example, Willis and Rosen (1979) with Murphy and Welch (1989). The former study analyzes data from the NBER–Thorndike Survey, estimates returns to schooling conditional on measured ability, and is explicitly concerned with the effect of unmeasured ability on the selection of students into schooling. The latter piece analyzes data from the Current Population Surveys, which contain no ability measures, and implicitly assumes that the selection of students into schooling is unrelated to ability. If experts can vary so widely in the way they infer the returns to schooling, it is reasonable to suspect that youth do as well.

Information and Incentive Policies for Youth at Risk

Our lack of understanding of the way youth perceive the returns to schooling figures prominently in the ongoing policy debate about the best way to influence the behavior of those youth considered at risk of dropping out of high school.

One line of thinking begins from the premise that remaining in school is generally in students' self-interest. If so, youth who drop out of school must misperceive the returns to schooling. Hence we should provide these youth with information that convinces them of the value of schooling. We should correct the misperceptions that presently lead youth to make poor schooling decisions.[8]

A second line of thinking begins from the premise that students perceive the returns to schooling correctly. If students choose to drop out, it is because they know that the returns to schooling are low.[9] This perspective implies that the key to improving school performance is better incentives. We should make sure that students who stay in school are rewarded and that those who do not are sanctioned.[10]

In the absence of evidence on how youth actually perceive the returns to schooling, policymaking goes on in the dark. Programs embodying many combinations of information and incentives are continually proposed and implemented across the country, with no basis for judging their effectiveness.[11]

5.3. Analysis of Intentions Data

Suppose that one wants to extrapolate individual behavior but finds the assumptions of revealed preference analysis to be unpalatable. If the only data available are observed choices, then the only option is to invoke other assumptions that one finds more palatable. But new data collection may open up new inferential possibilities.

It is often possible to elicit from people statements describing how they would behave when facing various decision problems. For example, female respondents in the June 1987 Supplement to the Current Population Survey (CPS) were asked this question:

> Looking ahead, do you expect to have any (more) children?
> Yes   No   Uncertain
>
> (U.S. Bureau of the Census, 1988a)

Respondents in the National Longitudinal Study of the High School Class of 1972 (NLS72) were asked these questions in fall 1973:

What do you expect to be doing in October 1974?

	(Circle one number on each line)	
	Expect to be doing	Do not expect to be doing
Working for pay at a full-time or part-time job	1	2
Taking vocational or technical courses at any kind of school or college	1	2
Taking academic courses at a two-year or four-year college	1	2
On active duty in the Armed Forces (or service academy)	1	2
Homemaker	1	2

(Riccobono et al., 1981)

Social scientists have long asked survey respondents to answer such *unconditional intentions* questions, and have used the responses to predict actual behavior. Responses to fertility questions like that in the CPS have been used for over fifty years to predict fertility (see Hendershot and Placek, 1981). Data on voting intentions have been used to predict American election outcomes since the early 1900s (see Turner and Martin, 1984). Surveys of buying intentions have been used to predict consumer purchase behavior since at least the mid-1940s (see Juster, 1964).

Social scientists also sometimes ask survey respondents to answer *conditional intentions* questions posing hypothetical scenarios. For example, a conditional-intentions version of the CPS fertility question might be

Imagine that the government were to enact a child-allowance program providing families with fifty dollars per month for each dependent child. Assuming that this program were in operation, would you expect to have any (more) children?

Yes   No   Uncertain

Interpretations of Intentions

The long history of intentions surveys notwithstanding, social scientists continue to differ in their interpretation of stated intentions and in the use they make of these data. Just as revealed preference analysis requires assumptions about the structure of preferences, interpretation of stated intentions requires assumptions about how people respond to the questions posed and how they actually behave. Social scientists disagree on these matters and so disagree on the interpretation of intentions data. It is particularly intriguing to contrast the perspectives of social psychologists and economists.

Social psychologists take intentions very seriously. They suppose that intention is a mental state that causally precedes behavior and that can be elicited through questionnaires or interviews. The influential work of Fishbein and Ajzen (1975) makes intention the intermediate variable in a behavioral model wherein (1) intentions are determined by attitudes and social norms and (2) behavior is determined by intentions alone.

According to Ajzen and Fishbein (1980), a person's *behavioral intention* is his subjective probability that the behavior of interest will occur. (They refer to the response to a yes/no intentions question as *choice intention*.) It seems, however, that social psychologists do not use the term "subjective probability" as a statistician would. Ajzen and Fishbein (1980, p. 50) state "we are claiming that intentions should always predict behavior, provided that the measure of intention corresponds to the behavioral criterion and that the intention has not changed prior to performance of the behavior." In a review of attitudinal research, Schuman and Johnson (1976, p. 172) write that the Fishbein-Ajzen model implies that "the correlation between behavioral intention and behavior should approach 1.0, provided that the focal behavior is the same in both cases and that nothing intervenes to alter the intention." It is difficult to reconcile these statements with the idea that behavioral intention is a subjective probability, unless that probability is always zero or one.

In practice, social psychologists typically measure intention on some nominal scale (Ajzen and Fishbein, 1980, for example, recommend a seven-point scale for which the verbally described responses range from "likely" to "unlikely") and report the arithmetic correlation between this measure and the behavioral outcome. See, for example, Schuman and Johnson (1976) and Davidson and Jaccard (1979).

In recent times, economists have mostly ignored intentions data. The dominant view is deep skepticism about the credibility of subjective statements of any kind. Early in their careers, economists are taught to believe only what people do, not what they say. Economists often assert that respondents to surveys have no incentive to answer questions carefully or honestly; hence, there is no reason to believe that subjective responses reliably reflect respondents' thinking.[12]

The economics discipline has not always been so hostile to the use of intentions data to predict behavior. From the mid-1950s through the mid-1960s, analysis of consumer buying intentions was close to a mainstream activity. Juster (1964, 1966) reviewed this literature and made original contributions of interest.

Considering the problem in which the behavior of interest is a binary purchase decision (buy or not buy) and the intentions question is also binary (intend to buy or do not intend to buy), Juster (1966, p. 664) wrote "Consumers reporting that they 'intend to buy A within X months' can be thought of as saying that the probability of their purchasing A within X months is high enough so that some form of 'yes' answer is more accurate than a 'no' answer." Thus he hypothesized that a consumer facing an intentions question responds as would a statistician asked to make a point prediction of a future event.

Perhaps intentions data reveal a mental state that causally precedes behavior. Perhaps they provide statistical predictions of behavior. Or perhaps *Webster's Eighth New Collegiate Dictionary* (1985, p. 629) is accurate when it states: "*Intention* implies little more than what one has in mind to do or bring about."

In the absence of consensus about the interpretation of stated intentions, it is of some use to place bounds on the predictive power of these data. The lower bound is necessarily zero. We cannot exclude the possibility that stated intentions reveal nothing about future behavior. To determine an upper bound, we might consider this thought experiment:

> Imagine ideal survey respondents who are fully aware of the process determining their behavior. What are the most informative responses that such ideal respondents can provide to intentions questions?

We shall study this thought experiment in the particularly simple context of intentions questions calling for yes/no predictions of binary outcomes (see also Manski, 1990b).

Rational-Expectations Responses to Intentions Questions

Let i and y be binary variables denoting the survey response and subsequent behavior respectively. Thus $i = 1$ if a person responds "yes" to the intentions question, and $y = 1$ if behavior turns out to satisfy the property of interest.

Suppose that a researcher observes a random sample of intentions responses i and respondent attributes x and is thus able to infer the distribution $P(i, x)$. The researcher, who does not observe respondents' subsequent behavior y, wishes to learn the choice probabilities $P(y \mid x, i)$ and $P(y \mid x)$; the former choice probability conditions on stated intentions and the latter does not. The inferential question is: What does knowledge of $P(i, x)$ reveal about $P(y \mid x, i)$ and $P(y \mid x)$?

I address this question under the assumption that (1) survey respondents are aware of the actual process determining their future behavior; and (2) given the information they possess at the time of the survey, respondents offer their best predictions of their behavior, in the sense of minimizing expected loss (see Section 1.2). When these conditions hold, intentions data are said to provide *rational-expectations* predictions of future behavior. The term *rational expectations* is routinely used by economists but should not be confused with the unrelated concept of rational choice.

A person giving a rational-expectations response to an intentions question would begin by recognizing that future behavior will depend in part on conditions known at the time of the survey and in part on events that have not yet occurred. A woman responding to the CPS fertility question, for example, should recognize that her future childbearing will depend not only on her current family and work conditions, which are known to her at the time of the survey, but also on the future evolution of these conditions, which she cannot predict with certainty.

Let s denote the information possessed by a respondent at the time that an intentions question is posed. (In the case of a conditional intentions question, s includes the information provided in the statement of the hypothetical scenario.) Let u represent uncertainty that will be resolved between the time of the survey and the time at which the behavior y is determined. Then y is necessarily a function of (s, u), and so may be written $y(s, u)$. I shall not impose any restrictions on the form of the function $y(s, u)$. In particular, behavior need not be determined by a rational-choice process.

Let $P_u \mid s$ denote the actual probability distribution of u conditional on s. Let $P(y \mid s)$ denote the actual distribution of y conditional on s. The event $y = 1$ occurs if and only if the realization of u is such that $y(s, u) = 1$. Hence, conditioning on s, the probability that $y = 1$ is

(5.8) $P(y = 1 \mid s) = P_u[y(s, u) = 1 \mid s]$.

A respondent with rational expectations is assumed to know $y(s, \cdot)$ and $P_u \mid s$ at the time of the survey; hence she knows $P(y = 1 \mid s)$.

The respondent is assumed to give her best point prediction of her future behavior, in the sense of minimizing expected loss. The best prediction necessarily depends on the losses the respondent associates with the two possible prediction errors, namely, $(i = 0, y = 1)$ and $(i = 1, y = 0)$. Whatever the loss function, however, the best prediction must satisfy the condition

(5.9) $i = 1 \Rightarrow P(y = 1 \mid s) \geq \pi$

$\qquad\;\; i = 0 \Rightarrow P(y = 1 \mid s) \leq \pi$,

for some threshold value $\pi \in (0, 1)$ that depends on the loss function. This formalizes the idea stated by Juster (1966). An alternative statement of (5.9) that will be used below is[13]

(5.9′) $P(y = 1 \mid s, i = 0) \leq \pi \leq P(y = 1 \mid s, i = 1)$.

Prediction of Behavior Conditional on Intentions

Consider a researcher who wishes to predict a survey respondent's future behavior y, conditional on her stated intention i and some attributes x observed by the researcher. That is, the researcher wishes to determine $P(y \mid x, i)$. The researcher does not know the stochastic process $P_u \mid s$ generating future events or the function $y(s, u)$ determining behavior.

A researcher who only knows that intentions are rational-expectations predictions can draw no conclusions about $P(y \mid x, i)$. There are two reasons. First, if the respondent's threshold value π is unknown, equation (5.9′) imposes no restrictions on $P(y = 1 \mid s, i)$. Second, if the information s possessed by the respondent is unknown, knowledge of $P(y = 1 \mid s, i)$ implies no restrictions on $P(y = 1 \mid x, i)$.

Conclusions can be drawn if

(a) the researcher knows π and includes π among the variables x, and

(b) respondents know the attributes x, so s subsumes x.

Then stated intentions reveal whether $P(y = 1 \mid x, i)$ is below or above π. To see this, observe that if condition (b) holds, the law of iterated expectations implies that

$$(5.10) \quad P(y = 1 \mid x, i) = E[P(y = 1 \mid s, i) \mid x, i].$$

If condition (a) holds, all persons with the same value of x share the same threshold value π. So (5.9′) and (5.10) together imply that

$$(5.11) \quad P(y = 1 \mid x, i = 0) \leq \pi \leq P(y = 1 \mid x, i = 1).$$

Prediction Not Conditioning on Intentions

Researchers often want to predict the behavior of nonsampled members of the population from which the survey respondents were drawn. Intentions data are available only for the sampled persons, so the predictions cannot now condition on i. Instead, the quantity of interest is $P(y = 1 \mid x)$.

Given conditions (a) and (b), the bound (5.11) obtained on $P(y = 1 \mid x, i)$ implies a bound on $P(y = 1 \mid x)$. Observe that

$$(5.12) \quad P(y = 1 \mid x) = P(y = 1 \mid x, i = 0) P(i = 0 \mid x)$$
$$+ P(y = 1 \mid x, i = 1) P(i = 1 \mid x).$$

Random sampling identifies the intentions probabilities $P(i = 0 \mid x)$ and $P(i = 1 \mid x)$, but we only know that the choice probabilities $P(y = 1 \mid x, i = 0)$ and $P(y = 1 \mid x, i = 1)$ lie in the intervals $[0, \pi]$ and $[\pi, 1]$, respectively. Hence (5.12) implies this sharp bound on $P(y = 1 \mid x)$:

$$(5.13) \quad \pi P(i = 1 \mid x) \le P(y = 1 \mid x)$$
$$\le \pi P(i = 0 \mid x) + P(i = 1 \mid x).$$

Observe that the bound width is $\pi P(i = 0 \mid x) + (1 - \pi) P(i = 1 \mid x)$. If $\pi = 1/2$, the bound width is $1/2$ for all values of $P(i \mid x)$.

It has been known for more than twenty-five years that intentions probabilities need not equal choice probabilities. That is, the relationship

$$(5.14) \quad P(i = 1 \mid x) = P(y = 1 \mid x)$$

need not hold (see Juster, 1966, p. 665). Nevertheless, some of the literature has considered deviations from this equality as "inconsistencies" in need of explanation. For example, Westoff and Ryder (1977, p. 449) state: "The question with which we began this work was whether reproductive intentions are useful for prediction. The basic finding was that 40.5 percent intended more, as of the end of 1970, and 34.0 percent had more in the subsequent five years. . . . In other words, acceptance of 1970 intentions at face value would have led to a substantial overshooting of the ultimate outcome." That is, the authors found that $P(i = 1 \mid x) = .405$, and subsequent data collection showed that $P(y = 1 \mid x) = .340$. Seeking to explain the observed "overshooting," the authors state: "One interpretation of our finding would be that the respondents failed to anticipate the extent to which the times would be unpropitious for childbearing, that they made the understandable but frequently invalid assumption that the future would resemble the present—the same kind of forecasting error that demographers have often made." More recent demographic studies continue to presume that deviations from (5.14) require explanation. See, for example, Davidson and Beach (1981) and O'Connell and Rogers (1983).

Rational-expectations prediction does imply (5.14) in the special case where future behavior depends only on the information s available at the time of the survey. Then a survey respondent can predict her future behavior with certainty. So i must equal y. But (5.14) need not hold if events u not known at the time of the survey partially determine future behavior. A simple example makes the point forcefully.

Suppose that respondents to the CPS fertility question report that they expect to have more children when childbirth is more likely than not; that is, $\pi = 1/2$. Suppose that the actual probability of having more children is always .51; that is, $P(y = 1 \mid s) = .51$ for all s. Then all women report that they intend to have more children; that is, $P(i = 1 \mid s) = 1$ for all s. This very substantial divergence between intentions and choice probabilities is consistent with the hypothesis that women report rational-expectations predictions of their future fertility. By equation (5.13), the rational-expectations hypothesis implies only that $P(y = 1 \mid s)$ lies between .50 and 1.

Empirical Evidence from the NLS72

The NLS72 data offer an opportunity to illustrate the bounds just derived. The schooling/work intentions questions quoted at the beginning of the section were followed by behavior questions asked in the fall of 1974:

What were you doing the first week of October 1974?

(Circle as many as apply)

Working for pay at a full-time or part-
time job   1

Taking academic courses at a two- or
four-year college   2

Taking vocational or technical courses
   at any kind of school or college   3
On active duty in the Armed Forces
   (or service academy)   4
Homemaker   5
Temporary lay-off from work, looking
   for work, or waiting to report to
   work   6

(Riccobono et al., 1981)

These questions about behavior correspond closely, though not per-fectly, to the intentions questions asked a year earlier.[14]

Table 5.2 presents empirical findings that condition on the re-spondent's sex and on the assumption that respondents use the com-mon threshold value $\pi = 1/2$. The quantities $P(i = 1 \mid x)$, $P(y = 1 \mid x)$, and $P(y = 1 \mid x, i)$ are estimated by corresponding sample fre-quencies. For example, the estimate of $P(\text{work} = 1 \mid x = \text{male}, i = 0)$ is based on the 2546 males who said they did not expect to work; the fraction of this group who reported working a year later was .42.

Inspection of the table shows that the male and female responses to the "work" and "academic" questions satisfy the bounds. So do the male responses to the "military" question and the female responses to the "homemaker" question. The female responses to the "military" question satisfy the bounds except for a small violation of bound (5.11) by those stating $i = 1$.

On the other hand, the responses of both sexes to the "voc.-tech." question and the male responses to the "homemaker" question violate the bounds substantially. Respondents who say they expect to take voc.-tech. courses later do so only 16 or 15 percent of the time. Males who say they expect to be homemakers later report themselves as such only 9 percent of the time.

Probabilistic Intentions

The use of intentions data to predict behavior has been controversial. At least part of the controversy stems from the fact that researchers have expected too much correspondence between stated intentions

Table 5.2 Consistency of schooling-work behavior in October 1974 with intentions stated in fall 1973

| | Bound (5.11) | | | |
| | Males | | Females | |
Behavior	Number of observations	Estimate of $P(y = 1 \mid x, i)$	Number of observations	Estimate of $P(y = 1 \mid x, i)$
Work	$i = 1$ 7143	.80	7439	.71
	$i = 0$ 2546	.42	2688	.31
Voc.-Tech.	$i = 1$ 1929	.16	1706	.15
	$i = 0$ 6972	.03	7575	.03
Academic	$i = 1$ 4829	.67	4320	.68
	$i = 0$ 4509	.06	5320	.06
Military	$i = 1$ 966	.67	158	.42
	$i = 0$ 7951	.012	8938	.002
Homemaker	$i = 1$ 246	.09	3626	.64
	$i = 0$ 8464	.004	5753	.08

	Bound (5.13) for $\pi = 1/2$					
	Males			Females		
	Estimates of			Estimates of		
Behavior	$P(i = 1 \mid x)$	Bound	$P(y = 1 \mid x)$	$P(i = 1 \mid x)$	Bound	$P(y = 1 \mid x)$
Work	.74	[.37–.87]	.67	.73	[.36–.86]	.58
Voc.-Tech.	.22	[.11–.61]	.05	.18	[.09–.59]	.05
Academic	.52	[.26–.76]	.33	.45	[.22–.72]	.30
Military	.11	[.05–.55]	.08	.017	[.008–.508]	.009
Homemaker	.03	[.01–.51]	.005	.39	[.19–.69]	.27

Source: Manski (1990b), table 1.

Note: The number of observations is not the same across questions because some respondents did not answer some questions. For example, 9689 males (i.e., 7143 + 2546) answered the work intentions question while 8917 (i.e., 966 + 7951) answered the military question.

and subsequent behavior. Social psychologists have written that intentions and behavior should coincide. Demographers have written that individual-level divergences between intentions and behavior should average out in the aggregate, so that (5.14) holds. In reality both premises are flawed. Intentions and behavior may diverge substantially, both at the individual level and in the aggregate, whenever behavior depends on events not yet realized at the time of the survey. This is so even if intentions data provide the best predictions of behavior that can be made given the information available when the survey is performed.

We have seen that binary intentions data have no predictive power in the absence of information about the threshold values π that respondents use in forming their responses. Even with π known, binary intentions data at most imply the bounds (5.11) and (5.13). It would seem that a superior survey approach would be to ask respondents for probabilistic assessments of their future behavior. For example, female respondents to the CPS could be asked:

Looking ahead, what is the percent chance that you will have any (more) children?

If expectations are rational, elicitation of probabilistic intentions reveals $P(y = 1 \mid s)$, which expresses all that can be said about future behavior given the information in s. Even if expectations are not rational, probabilistic intentions data may have greater predictive power than do binary data.

Juster (1966) not only proposed elicitation of probabilistic intentions but carried out an empirical study of probabilistic buying intentions. Market researchers have subsequently performed a variety of such studies (see Morrison, 1979; Urban and Hauser, 1980; and Jamieson and Bass, 1989). Psychologists studying judgment and decision making often elicit probabilistic predictions (see Wallsten and Budescu, 1983). Nevertheless, the intentions questions posed on major surveys continue to be predominately like those on the CPS and NLS72.

6

Simultaneity

A central objective of the social sciences is to learn about the ways in which individuals interact with one another. The enormous body of research on social interactions ranges from economic analysis of the anonymous process by which markets determine prices to ethnographic study of the intensely personal relationships among family members.

The last two chapters of this book examine identification problems that arise when observations of equilibrium outcomes are used to analyze social interactions. Chapter 7 considers the *reflection problem* that arises when a researcher observes the equilibrium distribution of behavior in a population and wishes to learn how the average behavior in some group influences the behavior of the individuals in the group. The present chapter examines the *simultaneity problem* that arises when observations of market transactions are used to study the demand behavior of price-taking consumers or the supply behavior of price-taking (or quantity-taking) firms. Simultaneity also arises in the analysis of nonmarket social interactions, when observations of equilibrium outcomes of games are used to study the reaction functions of the players. So simultaneity is a concern not only of economics but of the social sciences more generally.

6.1. "The" Identification Problem in Econometrics

A classical problem of econometrics is to infer the structure of supply and demand from observations of equilibrium prices and quantities. The basic version of the problem supposes that there is a set of isolated markets for a given product, each market separated from the others in time or in space. Each market has a value for $[s(\cdot), d(\cdot), p, q, x]$. Here x denotes some covariates characterizing a market, q is the quantity of product transacted, and p is the unit price at which these transactions take place. The market demand function $d(\cdot)$ gives the quantity of product that price-taking consumers would purchase if price were set at any level; so $d(t)$ is the quantity demanded if price were set equal to t. The market supply function $s(\cdot)$ gives the quantity of product that price-taking firms would offer if price were set at any level; so $s(t)$ is the quantity supplied if price were set equal to t. The transaction (p, q) is assumed to be an equilibrium outcome; that is, price p makes q both the quantity demanded and the quantity supplied. Formally, (p, q) is assumed to satisfy simultaneously the two conditions

$$(6.1) \quad q = d(p)$$

and

$$(6.2) \quad q = s(p).$$

Markets vary in their values of $[s(\cdot), d(\cdot), p, q, x]$. This heterogeneity is expressed by treating supply, demand, transactions, and covariates as random variables with some distribution $P[s(\cdot), d(\cdot), p, q, x]$. Let $P[s(\cdot), d(\cdot) \mid x]$ denote the distribution of supply and demand functions among markets sharing the same covariates. Econometric analysis seeks to learn about $P[s(\cdot), d(\cdot) \mid x]$ when observations of (p, q, x) are obtained by some sampling process, such as random sampling of markets, that reveals $P(p, q, x)$. Knowledge of $P(p, q, x)$ does not suffice to identify $P[s(\cdot), d(\cdot) \mid x]$. This is the *simultaneity problem*.

Analysis of simultaneity was so central to the early development of econometrics that it was long common for econometricians to think of identification and simultaneity as synonymous. Many econometrics texts, even recent ones, discuss identification only in the context of simultaneity. Particularly revealing is the title chosen by Fisher (1966) for his monograph on the simultaneity problem. He titled the book *The Identification Problem in Econometrics* and justified this choice by writing in the preface (p. vii): "Because the simultaneous equation context is by far the most important one in which the identification problem is encountered, the treatment is restricted to that context." From today's perspective, Fisher's judgment of the preeminence of simultaneity among all identification problems seems strained. Nevertheless, simultaneity remains an important problem of econometrics.

6.2. The Linear Market Model

The problem of learning $P[s(\cdot), d(\cdot) \mid x]$ may be posed for any specification of the covariates x. Given such a specification, the inferences that may be drawn depend on what is known a priori about the structure of supply and demand. Econometric analysis has long centered on the linear model

$$(6.3a) \quad s(t) = \beta_1 t + x'\alpha_1 + u_1$$

$$(6.3b) \quad d(t) = \beta_2 t + x'\alpha_2 + u_2$$

$$(6.3c) \quad E(u_1, u_2 \mid x) = 0.$$

Study of this linear model was initiated in the 1920s and crystallized by the early 1950s (see Hood and Koopmans, 1953).

Equation (6.3a) states that supply $s(\cdot)$ is a linear function of price t, with the same slope β_1 in each market. The supply function varies across markets only in its intercept $x'\alpha_1 + u_1$. This intercept is itself linear in x, a K-dimensional covariate vector observed by the researcher, and in u_1, a scalar covariate expressing determinants of supply that are unobserved by the researcher.

Equation (6.3b) imposes analogous restrictions on demand. Taken together, (6.3a) and (6.3b) transform the problem of learning the distribution of supply and demand into one of learning the parameters $(\beta_1, \beta_2, \alpha_1, \alpha_2)$ and the distribution $P(u_1, u_2 \mid x)$ of unobserved covariates. Condition (6.3c) is prior information about $P(u_1, u_2 \mid x)$. The mean of (u_1, u_2) in markets with covariates x equals zero.

Equations (6.3a) through (6.3c) imply that the mean regressions on x of the supply and demand functions have the linear form

(6.4a) $E[s(t) \mid x] = \beta_1 t + x' \alpha_1$

(6.4b) $E[d(t) \mid x] = \beta_2 t + x' \alpha_2.$

That is, if price were set equal to any value t, then the mean quantity supplied in markets with covariates x would be $\beta_1 t + x' \alpha_1$ and the mean quantity demanded in these markets would be $\beta_2 t + x' \alpha_2$.

The Algebra of Identification

When the linear model (6.3) is combined with the equilibrium conditions (6.1) and (6.2), we find that the equilibrium market outcome is

(6.5a) $q = x' \pi_1 + (\beta_1 u_2 - \beta_2 u_1)/(\beta_1 - \beta_2)$

(6.5b) $p = x' \pi_2 + (u_2 - u_1)/(\beta_1 - \beta_2),$

where

(6.6a) $\pi_1 \equiv (\beta_1 \alpha_2 - \beta_2 \alpha_1)/(\beta_1 - \beta_2)$

(6.6b) $\pi_2 \equiv (\alpha_2 - \alpha_1)/(\beta_1 - \beta_2).$

Thus equilibrium quantities and prices are linear functions of (x, u_1, u_2), with parameters that are known functions of the supply and demand parameters $(\beta_1, \beta_2, \alpha_1, \alpha_2)$. Equations (6.5a) and (6.5b) are commonly called the *reduced form* of the linear market model.

It follows from equations (6.5a), (6.5b), and (6.3c) that the mean regressions of equilibrium quantity and price on x are

(6.7a) $E(q \mid x) = x' \pi_1$

(6.7b) $E(p \mid x) = x' \pi_2.$

The sampling process identifies $E(q \mid x)$ and $E(p \mid x)$. Hence the parameters π_1 and π_2 are identified provided only that the covariates x are not perfectly collinear.

The function (6.6) mapping the supply and demand parameters $(\beta_1, \alpha_1, \beta_2, \alpha_2)$ into the reduced-form parameters (π_1, π_2) is many-to-one rather than one-to-one. So identification of (π_1, π_2) does not suffice to identify $(\beta_1, \alpha_1, \beta_2, \alpha_2)$. But the latter parameters are identified if knowledge of (π_1, π_2) is combined with suitable prior restrictions on $(\beta_1, \alpha_1, \beta_2, \alpha_2)$.

By far the most common practice is to invoke exclusion restrictions asserting that some components of α_1 (or α_2) equal zero while the corresponding components of α_2 (or α_1) are nonzero. For example, suppose it is known that $\alpha_{1K} = 0$ but $\alpha_{2K} \neq 0$. That is, holding $x_1, \ldots, x_{K-1}$ fixed, the mean regression of supply on x does not vary with x_K but the mean regression of demand on x does vary with x_K. Then $\pi_{1K}/\pi_{2K} = \beta_1$, so β_1 is identified. Moreover, $\pi_{1k} - \beta_1\pi_{2k} = \alpha_{1k}$, for $k = 1, \ldots, K$, so α_1 is identified. Thus exclusion of a component of α_1 identifies the mean supply regression (6.4a). Similarly, exclusion of a component of α_2 identifies the mean demand regression (6.4b). Observe that these results are obtained without imposing the usual economic assumptions that supply functions are upward sloping (that is, $\beta_1 \geq 0$) and demand functions are downward sloping (that is, $\beta_2 \leq 0$).[1]

Suppose that one can exclude at least one component of α_1 and at least one (different) component of α_2. Then the entire distribution of supply and demand is identified. The exclusion restrictions identify $(\beta_1, \alpha_1, \beta_2, \alpha_2)$, so we need only to identify the distribution $P(u_1, u_2 \mid x)$ of unobserved covariates. Rewrite the reduced-form equations (6.5) in the form

(6.5a′) $\beta_1 u_2 - \beta_2 u_1 = (\beta_1 - \beta_2)(q - x'\pi_1)$

(6.5b′) $u_2 - u_1 = (\beta_1 - \beta_2)(p - x'\pi_2)$.

It follows that

(6.8a) $u_1 = (q - x'\pi_1) - \beta_1(p - x'\pi_2)$

(6.8b) $u_2 = (q - x'\pi_1) - \beta_2(p - x'\pi_2)$.

With $(\pi_1, \pi_2, \beta_1, \beta_2)$ known, we can determine u_1 and u_2 in each market where (p, q, x) is observed; so we effectively observe realizations of (u_1, u_2, x). Thus $P(u_1, u_2 \mid x)$ is identified.

Predicting the Impact of a Tax

Identification of the distribution of supply and demand is not necessary to predict transactions in markets whose outcomes are observable. The sampling process identifies $P(p, q \mid x)$, so there is no need to learn $P[s(\cdot), d(\cdot) \mid x]$. Economists invoke models of market equilibrium in order to predict market outcomes in new settings. See Marschak (1953).

A typical application is to predict the transactions that would occur if a tax were to be placed on each unit of a product. Suppose that a tax of level τ is introduced. The value of τ may be positive or negative. In the latter case τ is usually called a subsidy, but there is no need to make this semantic distinction.

Let t continue to denote the price received by firms, so $t + \tau$ is now the price paid by consumers. Then equations (6.3a) and (6.3c) of the linear market model remain unchanged, but the demand specification of equation (6.3b) changes to

$$(6.9) \quad d(t) = \beta_2(t + \tau) + x'\alpha_2 + u_2.$$

Inserting (6.3a) and (6.9) into the equilibrium conditions (6.1) and (6.2) now yields not (6.5) but rather

$$(6.10a) \quad q = x'\pi_1 + \beta_1\beta_2\tau/(\beta_1 - \beta_2) + (\beta_1 u_2 - \beta_2 u_1)/(\beta_1 - \beta_2)$$

$$(6.10b) \quad p = x'\pi_2 + \beta_2\tau/(\beta_1 - \beta_2) + (u_2 - u_1)/(\beta_1 - \beta_2).$$

Comparing (6.10) with (6.5) shows that the impact of the tax on market outcomes is to shift the distribution of equilibrium quantities and prices. The mean regression of (q, p) on x, previously given in (6.7), is now

$$(6.11a) \quad E(q \mid x) = x'\pi_1 + \beta_1\beta_2\tau/(\beta_1 - \beta_2)$$

$$(6.11b) \quad E(p \mid x) = x'\pi_2 + \beta_2\tau/(\beta_1 - \beta_2).$$

Thus observation of market outcomes in the absence of a tax does not suffice to predict outcomes following introduction of a tax. But these outcomes can be predicted if the slope parameters β_1 and β_2 of the supply and demand functions are identified.

6.3. Equilibrium in Games

Many social interactions, from divorce proceedings to union-management negotiations to superpower rivalry, can usefully be thought of as games. The basic two-person game imagines that each of two players must select an action. Player 1 chooses an action from some set of feasible choices and player 2 chooses an action from his set of possibilities.

It is common to assume that the players have *reaction functions* $r_1(\cdot)$ and $r_2(\cdot)$ specifying the action that each would choose as a function of the action chosen by the other. Thus $r_1(t_2)$ specifies the action that player 1 would choose if player 2 were to select action t_2. Similarly, $r_2(t_1)$ specifies the action that player 2 would choose if player 1 were to select action t_1. An equilibrium of the game is a pair of mutually consistent actions. That is, (y_1, y_2) is an equilibrium pair of actions if

$$(6.12) \quad y_1 = r_1(y_2)$$

and

$$(6.13) \quad y_2 = r_2(y_1).$$

The problem of interest is to learn the distribution $P[r_1(\cdot), r_2(\cdot) \mid x]$ of reaction functions across games with specified covariates x. The simultaneity problem arises when one attempts to infer $P[r_1(\cdot), r_2(\cdot) \mid x]$ from observations of equilibrium outcomes.

Econometric analysis of two-person games has centered on the linear model

$$(6.14a) \quad r_1(t_2) = \beta_1 t_2 + x'\alpha_1 + u_1$$

(6.14b) $r_2(t_1) = \beta_2 t_1 + x'\alpha_2 + u_2$

(6.14c) $E(u_1, u_2 \mid x) = 0,$

which is analogous to the linear market model (6.3). Conditions (6.12) and (6.13) imply that the equilibrium outcome is

(6.15a) $y_1 = x'\pi_1 + (\beta_1 u_2 + u_1)/(1 - \beta_1\beta_2)$

(6.15b) $y_2 = x'\pi_2 + (\beta_2 u_1 + u_2)/(1 - \beta_1\beta_2),$

where

(6.16a) $\pi_1 \equiv (\beta_1\alpha_2 + \alpha_1)/(1 - \beta_1\beta_2)$

(6.16b) $\pi_2 \equiv (\beta_2\alpha_1 + \alpha_2)/(1 - \beta_1\beta_2).$

Hence the mean regressions of (y_1, y_2) on x are

(6.17a) $E(y_1 \mid x) = x'\pi_1$

(6.17b) $E(y_2 \mid x) = x'\pi_2.$

Analysis of identification follows the same lines as in the linear market model. The sampling process identifies $E(y_1 \mid x)$ and $E(y_2 \mid x)$, so π_1 and π_2 are identified if the covariates x are not perfectly collinear. The function (6.16) mapping $(\beta_1, \alpha_1, \beta_2, \alpha_2)$ into (π_1, π_2) is many-to-one rather than one-to-one, so knowledge of π_1 and π_2 does not identify $(\beta_1, \alpha_1, \beta_2, \alpha_2)$. But the latter parameters are identified if exclusion restrictions can be invoked. For example, suppose it is known that $\alpha_{1K} = 0$ but that $\alpha_{2K} \neq 0$. Then $\pi_{1K}/\pi_{2K} = \beta_1$, so β_1 is identified. Moreover, $\pi_{1k} - \beta_1\pi_{2k} = \alpha_{1k}$, for $k = 1, \ldots, K$, so α_1 is identified.

Ehrlich, the Supreme Court, and the National Research Council

A classical problem in criminology is to learn the deterrent effect of sanctions on criminal behavior. In the early 1970s it became common for criminologists to analyze observed crime rates and sanction levels as equilibrium outcomes of two-person games, wherein criminals (player 1) choose a crime rate and society (player 2) chooses sanctions. Linear reaction functions of the form (6.14) were used to specify the crime rate that criminals would choose if sanctions were set at any level t_2 and the sanctions that society would choose if the crime rate were t_1. In this setting, the parameter β_1 measures the deterrent effect of sanctions, that is, the change in the crime rate that would occur if sanctions were set at different levels.

The simultaneity problem in inference on deterrence became a concern beyond the community of academic criminologists when the Solicitor General of the United States (Bork et al., 1974) argued to the Supreme Court that a study by Isaac Ehrlich provided empirical evidence on the deterrent effect of capital punishment. Ehrlich (1975) used annual data on murders and sanctions in the United States to estimate a "murder supply" function specifying the murder rate that would occur as a function of sanctions levels, including the risk of capital punishment faced by a convicted murderer. He concluded (1975, p. 398): "In fact, the empirical analysis suggests that on the average the tradeoff between the execution of an offender and the lives of potential victims it might have saved was of the order of 1 for 8 for the period 1933–1967 in the United States."

This finding, and its citation before the Supreme Court as evidence in support of capital punishment, generated considerable controversy. A constructive outcome was a series of critiques of the Ehrlich study by Passell and Taylor (1975), Bowers and Pierce (1975), and Klein, Forst, and Filatov (1978), among others. Moreover, a panel of the National Research Council (NRC) was established to investigate in depth the problem of inference on deterrence (Blumstein, Cohen, and Nagin, 1978).

The NRC Panel on Research on Deterrent and Incapacitative Effects focused much of its attention on the simultaneity problem and stressed the difficulty of finding plausible exclusion restrictions to identify deterrent effects. The panel also examined in depth two other identification problems presenting obstacles to inference on deterrence: error in measuring crime rates and confounding of deterrence and incapacitation. In all, the panel report is an exceptionally clearheaded portrayal of the difficulties inherent in the empirical study of deterrence.

Regarding the deterrent effect of capital punishment, the panel concluded (p. 62): "The current evidence on the deterrent effect of capital punishment is inadequate for drawing any substantive conclusion." Cautious scientific assessments of this sort are not unusual in NRC studies, and are usually followed by calls for more research. But this NRC panel went on to draw a more unusual conclusion (p. 63):

> In undertaking research on the deterrent effect of capital punishment, however, it should be recognized that the strong value content associated with decisions regarding capital punishment and the high risk associated with errors of commission make it likely that any policy use of scientific evidence on capital punishment will require extremely severe standards of proof. The nonexperimental research to which the study of the deterrent effects of capital punishment is necessarily limited almost certainly will be unable to meet those standards of proof. Thus, the Panel considers that research on this topic is not likely to produce findings that will or should have much influence on policymakers.

This conclusion is both admirable and distressing. It is admirable that a panel of distinguished social scientists was willing to declare that social science research likely cannot resolve a behavioral question of vital public concern. Powerful incentives induce many researchers to maintain strong assumptions in order to draw strong conclusions. These researchers rejected the temptation.

The conclusion is distressing for the vaccuum it leaves. If research on the deterrent effect of capital punishment "should not have much influence on policymakers," what should influence them? Should policymakers reach conclusions about deterrence based solely on their own observations and reasoning, unaided by research? If so, how should they cope with the inferential problems that stymie social scientists? Should they succumb to the temptation to maintain whatever assumptions are needed to reach conclusions? Or should they give no weight to deterrence and base policy only on their normative views of just punishment? Neither option is appealing.

6.4. Simultaneity with Downward-Sloping Demand

Analysis of the simultaneity problem has long been confined to the linear models described in the preceding sections and to certain extensions thereof. I have found that fresh insights emerge if one puts aside these models and examines the probabilistic structure of the identification problem posed by simultaneity.[2] The analysis in this section focuses on inference on demand, but applies as well to inference on supply and to inference on reaction functions in games.

Simultaneity Is Selection

Simultaneity is actually a problem of censored outcomes. To see this, let x be a specified point on the support of the covariate distribution and let t be a specified price. Consider the distribution $P[d(t) \mid x]$ of the quantity demanded at price t in markets with covariates x. Write

$$(6.18) \quad P[d(t) \mid x] = P[d(t) \mid x, p = t] P(p = t \mid x)$$
$$+ P[d(t) \mid x, p \neq t] P(p \neq t \mid x).$$

Assume that condition (6.1) holds, so the transaction in each market lies on that market's demand function. Then

$$(6.19) \quad P[d(t) \mid x, p = t] = P(q \mid x, p = t).$$

The sampling process reveals $P(p = t \mid x)$, $P(p \neq t \mid x)$, and $P(q \mid x, p = t)$, but does not reveal $P[d(t) \mid x, p \neq t]$. This is precisely the selection problem. The selection probability is $P(p = t \mid x)$. The censoring probability is $P(p \neq t \mid x)$. The distribution of outcomes conditional on selection is $P[d(t) \mid x, p = t]$, or $P(q \mid x, p = t)$ by (6.19). The distribution of outcomes conditional on censoring is $P[d(t) \mid x, p \neq t]$. One wishes to learn about $P[d(t) \mid x]$, the distribution of outcomes that would be observed if price were set equal to t in all markets with covariates x.

Suppose one knows that condition (6.1) holds but has no prior information restricting the structure of demand or supply. Then observations of outcomes in markets where $p \neq t$ reveal nothing about the censored demand distribution $P[d(t) \mid x, p \neq t]$. Thus the *worst-case* analysis of Section 2.2 applies.

In the presence of prior information, observations of outcomes in markets where $p \neq t$ may be informative about the censored demand distribution $P[d(t) \mid x, p \neq t]$. We saw in Section 6.2 that the linear market model accompanied by an exclusion restriction identifies $P[d(t) \mid x]$. Here I consider a different kind of prior information.

Ordered Outcomes

If an economist is willing to assume anything about the structure of demand, it generally is that demand is weakly downward sloping; that is,

(6.20) $t' > t \Rightarrow d(t') \leq d(t)$.

Yet the literature on simultaneity has not studied the identifying power of this most common economic assumption.

It is intuitive that the assumption of downward-sloping demand should have identifying power. Consider Figure 6.1. If we know only

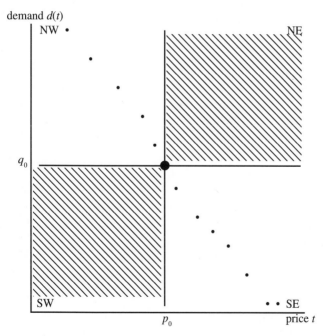

Figure 6.1   Feasible demand functions

that (6.1) holds, then observation of a market outcome (p_o, q_o) reveals only that $d(\cdot)$ is some function passing through the point (p_o, q_o). But if we know that (6.1) and (6.20) hold, then observation of (p_o, q_o) reveals that the downward-sloping $d(\cdot)$ must lie entirely within the northwest (NW) and southeast (SE) regions of the figure, as does the curve drawn.

In Chapter 2, we studied the simplest nontrivial case of assumption (6.20). Recall that the literature analyzing treatment effects supposes that a binary variable z determines which of two outcomes is observed; outcome y_1 is observed if $z = 1$ and outcome y_0 is observed if $z = 0$. In Section 2.6 we examined the identifying power of the *ordered outcomes* assumption[3]

$$(6.21) \quad y_1 \leq y_0.$$

Assumption (6.21) is the special case of (6.20) with $t = 0$ and $t' = 1$. Henceforth I use the term *ordered outcomes* to refer to assumption (6.20) and not just to the special case (6.21).[4]

Bounds on Conditional Probabilities

In Manski (1994c), I show that assumptions (6.1) and (6.20) imply bounds on various features of the distribution $P[d(t) \mid x]$. It is particularly simple to see that these assumptions imply bounds on the probability $P[d(t) \leq r \mid x]$ that the quantity demanded at price t is smaller than a specified constant r.

Figure 6.2 decomposes the possible market outcomes into four regions based on the position of (p, q) relative to (t, r). Given that demand is downward sloping, each observation of (p, q) in the SW region implies that $d(t) \leq r$ and each observation in the NE region implies that $d(t) > r$. Observations of (p, q) in the SE and NW regions do not reveal whether $d(t)$ is less than or greater than r. Formally, the following holds in each region:[5]

SW region: $p \leq t \cap q \leq r \Rightarrow d(t) \leq d(p) = q \leq r$

SE region: $p > t \cap q \leq r \Rightarrow d(t) \geq d(p) = q$

equilibrium quantity

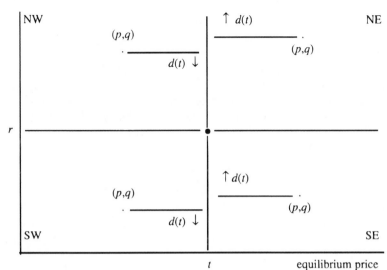

Figure 6.2   Identification of $P[d(t) \leq r \mid x]$.

NW region: $p < t \cap q > r \Rightarrow d(t) \leq d(p) = q$

NE region: $p \geq t \cap q > r \Rightarrow d(t) \geq d(p) = q > r.$

So we have this sharp bound on $P[d(t) \leq r \mid x]$:

$$(6.22) \quad P(p \leq t \cap q \leq r \mid x) \leq P[d(t) \leq r \mid x]$$
$$\leq 1 - P(p \geq t \cap q > r \mid x).$$

The width of the bound depends on the distribution of market outcomes. At one extreme, the support of $P(p, q \mid x)$ may be concentrated in the SE and NW regions of Figure 6.2. Then (6.22) becomes $0 \leq P[d(t) \leq r \mid x] \leq 1$. At the other extreme, the support of $P(p, q \mid x)$ may be concentrated in the SW and NE regions. Then (6.22) becomes $P[d(t) \leq r \mid x] = P(p \leq t \cap q \leq r \mid x)$. So the equilibrium condition (6.1) and the ordered outcome assumption (6.20) may reveal nothing about $P[d(t) \leq r \mid x]$ or may identify this quantity.

The Effect of Policing on Crime

Section 6.3 called attention to the problem of inference on the deterrent effect of sanctions on criminal behavior. I shall use this inferential problem to provide an empirical illustration of the bound (6.22).

The objective is to use observations of crime rates and deterrence policies to study the effect of deterrence policies on criminal behavior. The standard setup assumes a set of isolated jurisdictions. Each jurisdiction has a crime function $d(\cdot)$ giving the crime rate that would occur if deterrence (that is, the price of crime) were set at any level. Some process determines the actual deterrence level p in each jurisdiction. The realized crime rate is then $q = d(p)$. The problem is to learn about the distribution $P[d(\cdot) \mid x]$ of crime functions among jurisdictions with covariates x. Our interest is to learn what inferences are possible if it is assumed only that crime is a weakly decreasing function of the deterrence level.

Empirical studies have measured crime and deterrence in many different ways. To provide an accessible illustration, I use data for American cities with populations over 25,000 published in U.S. Census Bureau (1988b), *County and City Data Book 1988*. The crime rate is an FBI estimate of the number of serious crimes committed per 100,000 resident population in the year 1985 (table C, item 31). The deterrence measure is an FBI estimate of the number of police officers per 10,000 resident population in 1985 (table C, item 33). I use these crime rates and police densities as reported, except that I rescale the crime rate so that it has the same population base as the police density.

I focus on the twenty-two cities in the state of Wisconsin.[6] These observations form a data set large enough to yield interesting findings, but small enough to permit one to follow the calculations easily. Figure 6.3 presents the raw data and displays the configuration of police densities and crime rates.

Assume that the crime functions in these Wisconsin cities are a random sample of size $N = 22$ drawn from the distribution $P[d(\cdot) \mid x]$ of crime functions in jurisdictions sharing specified covariates x.[7] Then consistent estimates of the bound (6.22) may be obtained by replacing the probabilities defining the bound with the corresponding sample frequencies. Table 6.1 presents a set of such estimates. To keep the discussion centered on the problem of identification, I discuss the table as if its entries are the bounds rather than just estimates thereof.

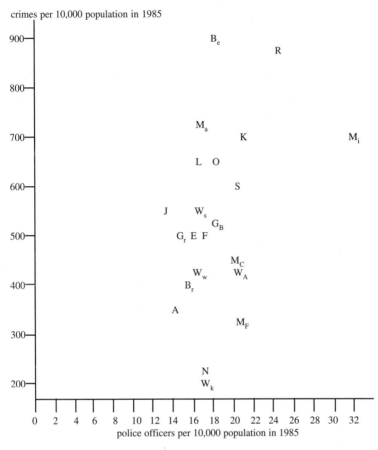

crimes per 10,000 population in 1985

police officers per 10,000 population in 1985

Key

A - Appleton (14.0, 379.2)
B$_e$ - Beloit (18.5, 906.0)
B$_r$ - Brookfield (15.7, 400.3)
E - Eau Claire (16.0, 493.9)
F - Fond du Lac (17.2, 487.2)
G$_B$ - Green Bay (18.2, 540.7)
G$_r$ - Greenfield (15.1, 508.4)
J - Janesville (13.4, 561.5)
K - Kenosha (19.8, 690.4)
L - La Crosse (17.2, 647.1)
M$_a$ - Madison (17.2, 722.5)

M$_c$ - Manitowoc (19.4, 458.2)
M$_F$ - Menomonee Falls (20.3, 341.1)
M$_i$ - Milwaukee (32.8, 706.6)
N - New Berlin (17.0, 222.6)
O - Oshkosh (17.3, 631.5)
R - Racine (24.1, 857.2)
S - Sheboygan (19.2, 601.3)
W$_k$ - Waukesha (17.1, 221.2)
W$_s$ - Wausau (16.5, 552.9)
W$_w$ - Wauwatosa (17.5, 450.8)
W$_A$ - West Allis (20.2, 425.6)

Figure 6.3   Crime rates and police densities in Wisconsin cities. (Data
        source: U.S. Bureau of the Census, *County and City Data Book
        1988,* table C.)

Table 6.1   Estimated bounds on features of $P[d(\cdot) \mid x]$

| | $P[d(t) \le r \mid x]$ | | | |
	$r = 200$	$r = 400$	$r = 600$	$r = 800$
$t = 12$	[.00, .00]	[.00, .18]	[.00, .64]	[.00, .91]
$t = 16$	[.00, .23]	[.05, .36]	[.23, .64]	[.23, .91]
$t = 20$	[.00, .82]	[.14, .86]	[.55, .91]	[.77, .95]
$t = 24$	[.00, .91]	[.18, .91]	[.64, .91]	[.86, .95]

The most striking feature is how much the estimated bounds vary in width as (t, r) varies. We are able to draw only very weak inferences when $(t, r) = (12, 800)$ and $(t, r) = (24, 200)$. In these cases, the table entries are $P[d(12) \le 800 \mid x] \in [.00, .91]$ and, likewise, $P[d(24) \le 200 \mid x] \in [.00, .91]$. We can, however, draw rather strong inferences when $(t, r) = (12, 200)$ and $(t, r) = (24, 800)$. Here $P[d(12) \le 200 \mid x] = 0$ and $P[d(24) \le 800 \mid x] \in [.86, .95]$. That is, taking the estimates at face value, we find that setting the police density equal to 12 officers per 10,000 population always yields a crime rate larger than 200 crimes per 10,000 population. Setting the police density equal to 24 officers per 10,000 population has at least a .86 chance of yielding a crime rate less than or equal to 800 crimes per 10,000 population.

7

The Reflection Problem

7.1. Endogenous, Contextual, and Correlated Effects

There is a long-standing interdisciplinary split between economists and sociologists on the channels through which society affects the individual. Whereas sociologists hypothesize that society affects individuals in myriad ways, economists often assume that society acts on individuals only by constraining their opportunities. Many economists regard such central sociological concepts as norms and reference groups as spurious epiphenomena explainable by processes operating entirely at the level of the individual. See, for example, the Friedman (1957) criticism of Duesenberry (1949).[1]

Even among sociologists, one does not find consensus on the nature of social interactions. Consider the ongoing debate about the meaningfulness of the concept of the underclass and the related controversy about the existence and nature of neighborhood effects (see Jencks and Mayer, 1989). Or consider the split between those sociologists who take class, ethnic group, or gender as the fundamental unit of analysis, and those who view society as a collection of heterogeneous individuals, families, and households.

I began this book by calling attention to two hypotheses often advanced to explain the common observation that individuals belonging to the same group tend to behave similarly. These hypotheses, plus a third not mentioned previously, are

endogenous effects, wherein the propensity of an individual to behave in some way varies with the prevalence of that behavior in the group;

contextual effects, wherein the propensity of an individual to behave in some way varies with the distribution of background characteristics in the group; and

correlated effects, wherein individuals in the same group tend to behave similarly because they face similar institutional environments or have similar individual characteristics.

Endogenous and contextual effects express distinct ways that individuals might be influenced by their social environments; correlated effects express a nonsocial phenomenon. Consider, for example, the high school achievement of a teenage youth. There is an endogenous effect if, all else equal, individual achievement tends to vary with the average achievement of the students in the youth's high school, ethnic group, or other reference group. There is a contextual effect if achievement tends to vary with, say, the socioeconomic composition of the reference group. There are correlated effects if youth in the same school tend to achieve similarly because they are taught by the same teachers or because they have similar family backgrounds.

Sociologists, social psychologists, and some economists have long been concerned with *reinforcing* endogenous effects, wherein the propensity of an individual to behave in some way increases with the prevalence of that behavior in the reference group. A host of terms are commonly used to describe these effects: "conformity," "imitation," "contagion," "bandwagon," "herd behavior," "norm effects," and "keeping up with the Joneses." Economists have always been fundamentally concerned with a particular nonreinforcing endogenous effect: an individual's demand for a product varies with price, which is partly determined by aggregate demand in the relevant market.

Contextual effects became an important concern of sociologists in the 1960s, when substantial efforts were made to learn how youth are influenced by their school and neighborhood environment (Coleman et al., 1966; Sewell and Armer, 1966; Hauser, 1970). The recent resurgence of interest in spatial concepts of the underclass has spawned new empirical studies (for example, Crane, 1991; Jencks and Mayer, 1989; and Mayer, 1991).

Distinguishing among endogenous, contextual, and correlated effects is important because these hypotheses have differing implications for the prediction of social interactions. For example, consider an educational intervention providing tutoring to some of the students in a school but not to the others. If individual achievement increases with the average achievement of the students in the school, then an effective tutoring program not only directly helps the tutored students but, as their achievement rises, indirectly helps all students in the school, with a feedback to further achievement gains by the tutored students. Contextual and correlated effects do not generate this *social multiplier*.

Although endogenous and contextual effects differ conceptually and in their predictive implications, these two types of social effect have often been confused. For example, studies of school integration, typified by Coleman et al. (1966), seem to have in mind an endogenous social effect, wherein the achievement of each student is affected by the mean achievement of the students in the same school. But these studies generally estimate contextual-effects models, wherein the achievement of each student is affected by the racial composition of his school.

The same tension appears in recent analyses of neighborhood effects. The theoretical section of Crane (1991) poses an "epidemic" model of endogenous neighborhood effects, wherein a teenager's school dropout and childbearing behavior is influenced by the neighborhood frequency of dropping out and childbearing. But Crane estimates a contextual-effects model, wherein a teenager's behavior depends on the occupational composition of her neighborhood. This juxtaposition of endogenous-effect theorizing and contextual-effect empirical analysis also appears in Jencks and Mayer (1989).

This chapter analyzes the *reflection* problem that arises when a researcher observes the distribution of behavior in a population and wishes to infer whether the average behavior in some group influences the behavior of the individuals that compose the group.[2] I refer to this as a reflection problem because it is similar to the problem of interpreting the almost simultaneous movements of a person and his reflection in a mirror. Does the mirror image cause the person's movements or reflect them? An observer who does not understand something of optics and human behavior would not be able to tell.

7.2. A Linear Model

The essence of the reflection problem can be seen through examination of a linear model often invoked in empirical studies. This model characterizes each member of the population by a value for the variables (y, x, w, u). Here y is a scalar outcome, x are variables characterizing an individual's reference group, and (w, u) are other variables that directly affect y. For example, y might be a youth's achievement in high school, x might denote his or her ethnic group, and (w, u) might be the youth's socioeconomic status and ability.

A researcher observes a random sample of (y, x, w) but does not observe the associated realizations of u. The variables (x, w, u) are known to determine y through a linear model

(7.1a) $y = \alpha + \beta E(y \mid x) + E(w \mid x)'\gamma + w'\lambda + u$

(7.1b) $E(u \mid x, w) = x'\delta,$

where $(\alpha, \beta, \gamma, \delta, \lambda)$ is a parameter vector. Then the mean regression of y on (x, w) has the linear form

(7.2) $E(y \mid x, w) = \alpha + \beta E(y \mid x) + E(w \mid x)'\gamma + x'\delta + w'\lambda.$

If $\beta \neq 0$, the linear regression (7.2) expresses an endogenous social effect: the response y varies with $E(y \mid x)$, the mean of the endogenous variable y among those persons in the reference group described by x.[3] If $\gamma \neq 0$, the model expresses a contextual effect: y varies with $E(w \mid x)$, the mean of the exogenous variables w among those persons in the reference group. If $\delta \neq 0$, the model expresses correlated effects: persons in reference group x tend to have similar unobserved attributes u. The parameter λ expresses the direct effect of w on y.

Parameters as Treatment Effects

What legitimates use of the term "effect" to describe the parameters $(\beta, \gamma, \delta, \lambda)$? The term is appropriate if values of $[E(y \mid x), E(w \mid x), x, w]$ are treatments assigned exogenously to individuals, in the manner of Section 2.5. Then the parameters are classical treatment effects measuring the changes in expected outcome associated with a unit change in each component of $[E(y \mid x), E(w \mid x), x, w]$, holding the other components fixed.

The sample data alone cannot identify these treatment effects. The variables $E(y \mid x)$, $E(w \mid x)$, and x are functionally dependent in the population under observation; they are all functions only of x. Data drawn from this population cannot reveal what outcomes would occur if the empirical relationship among these three variables were to be altered. The social-effects model (7.2), if identified, makes it possible to predict outcomes under all logically possible configurations of the variables $[E(y \mid x), E(w \mid x), x, w]$. The model also makes it possible to predict outcomes if the parameters $(\alpha, \beta, \gamma, \delta, \lambda)$ were to change in specified ways.

Identification of the Parameters

Suppose that (x, w) has discrete support or, alternatively, that $E(y \mid x, w)$, $E(y \mid x)$, and $E(w \mid x)$ vary continuously with (x, w). Then random sample data on (y, x, w) identifies these regressions on the support of (x, w) (see Section 1.3).[4] At each value of (x, w), whether on the support or off, the social-effects representation of $E(y \mid x, w)$ on the right side of (7.2) is a function of the parameters $(\alpha, \beta, \gamma, \delta, \lambda)$. So the inferential question is: What does knowledge of $E(y \mid x, w)$, $E(y \mid x)$, $E(w \mid x)$, x, and w imply about $(\alpha, \beta, \gamma, \delta, \lambda)$?

The reflection problem arises out of the presence of $E(y \mid x)$ as a regressor in (7.2). Taking expectations of both sides of (7.2) with respect to w reveals that $E(y \mid x)$ solves the *social equilibrium* equation

(7.3) $E(y \mid x) = \alpha + \beta E(y \mid x) + E(w \mid x)'\gamma + x'\delta + E(w \mid x)'\lambda.$

Provided that $\beta \neq 1$, equation (7.3) has a unique solution, namely,

(7.4) $E(y \mid x) = [\alpha + E(w \mid x)'(\gamma + \lambda) + x'\delta]/(1 - \beta).$

Thus model (7.2) implies that $E(y \mid x)$ is a linear function of $[1, E(w \mid x), x]$, where "1" denotes the constant. In other words, $E(y \mid x)$ is *perfectly collinear* with $[1, E(w \mid x), x]$.

As is well known, perfect collinearity of the regressors in a linear model implies that the associated parameters are unidentified, which means that here α, β, γ, and δ are unidentified. Endogenous effects

of the behaviors of the other sample members. Thus the model assumes that an endogenous effect is generated within the researcher's sample rather than within the population from which the sample was drawn. This makes sense in studies of small-group interactions, where the sample is composed of clusters of friends, co-workers, or household members; see, for example, Duncan, Haller, and Portes (1968, 1970) or Erbring and Young (1979). But it does not make sense in studies of neighborhood and other large-group social effects (for example, Case, 1991), where the sample members are randomly chosen individuals.

There is a way to make sense of the spatial correlation model in studies of large-group interactions. This model can be interpreted as a two-stage method for estimating a pure endogenous-effects model. In the first stage, one uses data on (y, x) to estimate $E(y \mid x)$ nonparametrically, and in the second stage, one estimates (α, β, λ) by finding the least squares fit of y to $[1, E_N(y \mid x), w]$, where $E_N(y \mid x)$ is the first-stage estimate of $E(y \mid x)$. Many nonparametric estimates of $E(y \mid x_i)$, including the local average of Section 1.3, are weighted averages of the form $E_N(y \mid x_i) = H_{iN}Y$, with H_{iN} determining the specific estimate. Hence estimates of (α, β, λ) reported in the literature on spatial correlation can be interpreted as estimates of pure endogenous-effects models.

7.4. Inferring the Composition of Reference Groups

Researchers studying social effects rarely offer empirical evidence to support their specifications of reference groups. The prevailing practice is simply to assume that individuals are influenced by $E(y \mid x)$ and $E(w \mid x)$, for some specified x. Among the few studies that do bring empirical evidence to bear are Coleman, Katz, and Menzel (1957), Duncan, Haller, and Portes (1968, 1970), and Woittiez and Kapteyn (1991). These studies attempt to elicit reference-group information from survey respondents.

Suppose that a researcher does not know the composition of reference groups. Can the researcher infer the composition of reference groups from the observed distribution of behavior? The answer is negative.

To see this, let x be any hypothesized specification of reference groups and let w be any specified function of x, say, $w = w(x)$. Knowledge of x implies knowledge of $w(x)$. Hence the equality

(7.9) $E[y \mid x, w(x)] = E(y \mid x)$

holds tautologically. Comparing equations (7.9) and (7.2) shows that (7.9) is the special case of (7.2) in which $\beta = 1$ and $\alpha = \gamma = \delta = \lambda = 0$. Thus the observed distribution of behavior is always consistent with the hypothesis that individual behavior reflects the average behavior of reference group x.

For example, suppose that a researcher studying student achievement specifies $x =$ (ability, family income) and $w =$ ability. The researcher will find that the observed distribution of behavior is always consistent with this hypothesis: individual achievement varies directly with the average achievement of the specified reference group but does not vary with the student's own ability.

7.5. Dynamic Analysis

We have seen that the reflection problem can make it rather difficult to draw conclusions about the nature of social effects from observations of the equilibrium outcomes experienced by a population. Informed specification of reference groups is necessary before empirical analysis can be contemplated. Even then, endogenous effects cannot be distinguished from contextual or from correlated effects.

Observation of the dynamics of social processes can sometimes open new possibilities for inference. Consider this dynamic version of the linear model (7.2):

(7.10) $E_t(y \mid x, w) = \alpha + \beta E_{t-1}(y \mid x)$
$$+ E_{t-1}(w \mid x)'\gamma + x_t'\delta + w_t'\lambda,$$

where E_t and E_{t-1} denote expectations taken at periods t and $t - 1$. The idea is that nonsocial forces act contemporaneously but social forces act on the individual with a lag. Empirical studies based on (7.10) include Alessie and Kapteyn (1991) and Borjas (1992).

The dynamic linear model (7.10) may have a unique stable temporal equilibrium of the form (7.3). In particular, this is the case if $-1 < \beta < 1$ and the variables $E(w \mid x)$ and x are time-invariant. When equilibrium outcomes are observed, the reflection problem arises just as in the static model treated earlier. But the reflection problem does not arise when nonequilibrium outcomes are observed. Out of equilibrium, $E_{t-1}(y \mid x)$ is not necessarily a linear function of $[1, E_{t-1}(w \mid x), x_t]$.

Of course one cannot simply specify a dynamic model and claim that the problem of inference on social effects has been resolved. Dynamic analysis is meaningful only if one has reason to believe that the transmission of social effects follows the assumed temporal pattern. Are individuals influenced by the behavior of their contemporaries or by the experiences of those a few years older? Or do social effects operate across generations? As matters stand, we do not know.

Notes

Introduction

1. Intrusiveness is precluded not only for ethical reasons but also because behavior may change when people know they are being observed.
2. Stigler (1986) provides an interesting account of the early history.

1. Extrapolation

1. The notation $P(y \mid x)$ is variously used in this book to denote

   - the distribution of y conditional on x, viewed as a function of x
   - the distribution of y conditional on x, evaluated at a particular value of x
   - the probability that y takes a particular value conditional on x, viewed as a function of x
   - the probability that y takes a particular value conditional on x, evaluated at a particular value of x.

   The correct interpretation is usually clear from the context. Where there seems some chance of ambiguity, I make the correct interpretation explicit.
2. If the minimization problem has multiple solutions, then these are all best predictors of y conditional on x. It is sometimes convenient, though not necessary, to use an auxiliary rule to pick out one solution.
3. A useful generalization of absolute loss, treating under- and over-

predictions of y asymmetrically, is the asymmetric absolute loss function

$$L(u) = (1 - \alpha) |u| \quad \text{if } u \leq 0$$
$$= \quad \alpha |u| \quad \text{if } u \geq 0,$$

where α is a specified constant in the interval $(0, 1)$. The resulting best predictor is the α-quantile of y conditional on x; that is, the smallest number t such that $P(y \leq t \mid x) \geq \alpha$. This somewhat cumbersome definition of a quantile simplifies if y has a smoothly increasing distribution function; then the α-quantile is the unique value t such that $P(y \leq t \mid x) = \alpha$. The median is the name given to the 0.5 quantile. See Ferguson (1967) and Manski (1988a).

4. Throughout this book, the term "prior information" is used in the classical sense to refer to restrictions that are known with certainty to be satisfied by the population of interest. The literature on Bayesian inference uses the term in a more general sense to refer to restrictions that may or may not be satisfied by the population; the researcher places a subjective probability on the event that the restriction holds. Beginning from subjectively probabilistic assumptions, Bayesian inference generates subjectively probabilistic conclusions about identification.

5. The literature on nonparametric regression refers to (1.3) as a *uniform kernel* estimate. The reader wishing to learn more about this and other nonparametric regression methods may turn to Manski (1991) and then, for more depth, to Hardle (1990).

   To understand why conditions (a), (b), and (c) suffice, suppose that d_N is kept fixed at some value d. Then as the sample size increases, the strong law of large numbers implies that the estimate (1.3) converges almost surely to the probability that y falls in B, conditional on the event that x falls within a distance d of x_0; that is, to $P[y \in B \mid (|x - x_0|) < d]$. If $P(y \in B \mid x)$ is continuous near x_0, then as d approaches zero, $P[y \in B \mid (|x - x_0|) < d]$ approaches $P(y \in B \mid x = x_0)$. These two facts suggest that an estimate converging to $P(y \in B \mid x = x_0)$ can be obtained by letting the bandwidth d_N approach zero as N increases. This heuristic idea succeeds provided that d_N does not approach zero too quickly as the sample size grows. In particular, it turns out that the rate at which d_N approaches zero must be slower than $1/N^{1/K}$, where K is the dimension of the vector x. This condition ensures that the number of observations actually used to calculate the local-frequency estimate (1.3) increases with the sample size.

Actually this is a special case of a much more general result. Let $g(y)$ be any function of y and consider the conditional expectation $E[g(y) \mid x = x_0]$. This may be estimated by the local average

$$\frac{\sum_{i=1}^{N} g(y_i) \, 1[\lvert x_i - x_0 \rvert < d_N]}{\sum_{i=1}^{N} 1[\lvert x_i - x_0 \rvert < d_N]} \, .$$

This estimate almost surely converges to $E[g(y) \mid x = x_0]$ as the sample grows in size, provided that these conditions hold:

(a1) $E[g(y) \mid x]$ varies continuously with x, for x near x_0.

(a2) The conditional variance $V[g(y) \mid x]$ is bounded for x near x_0.

(b) One tightens the bandwidth as the sample size increases.

(c) One does not tighten the bandwidth too rapidly.

Condition (a2) is automatically satisfied when $g(y)$ is a bounded function, as is the indicator function $1[y \in B]$.

6. A remark on statistical inference is warranted here. When $P(x = x_0) > 0$, the frequency estimate (1.2) is the most precise possible in the absence of prior information restricting $P(y \mid x)$. When $P(x = x_0) = 0$ but x_0 is on the support, the theory of optimal estimation of $P(y \mid x)$ is much more involved. In general, the local frequency estimate (1.3) is not the most precise possible. See Hardle (1990).

7. For example, differentiability conveys a stronger sense of smoothness than does continuity, but this assumption too only restricts the behavior of $P(y \mid x)$ near x_1.

8. Nonparametric estimates of the sampling distributions for these estimates are reported in table 5 of Manski et al. (1992). Ninety-percent confidence intervals for the estimates are as follows:

90% Confidence intervals for the estimates in Table 1.1

	Intact family	Nonintact family
White male	[.86, .92]	[.68, .86]
White female	[.91, .96]	[.69, .88]
Black male	[.80, .91]	[.72, .85]
Black female	[.90, .98]	[.80, .92]

2. The Selection Problem

1. The material in this chapter is drawn largely from Manski (1989, 1990a, 1994a) and from Manski et al. (1992).

2. It would also be appropriate to describe (2.3) as an invariance assumption of the kind discussed in Section 1.4 in the context of extrapolation.

3. A more general form of the bound (2.6) can be shown for the mean of a bounded function of y. Let $g(y)$ be a function mapping y into some known bounded interval $[K_0, K_1]$. Observe that

$$E[g(y) \mid x] = E[g(y) \mid x, z = 1] P(z = 1 \mid x)$$
$$+ E[g(y) \mid x, z = 0] P(z = 0 \mid x).$$

The sampling process identifies $E[g(y) \mid x, z = 1]$ and $P(z \mid x)$ but provides no information on $E[g(y) \mid x, z = 0]$. The last quantity, however, necessarily lies in the interval $[K_0, K_1]$. Hence,

$$E[g(y) \mid x, z = 1] P(z = 1 \mid x) + K_0 P(z = 0 \mid x)$$
$$\leq E[g(y) \mid x]$$
$$\leq E[g(y) \mid x, z = 1] P(z = 1 \mid x) + K_1 P(z = 0 \mid x).$$

Thus a censored-sampling process bounds the conditional expectation of any bounded function of y. The bound (2.6) emerges if one lets $g(y)$ be the indicator function $1[y \in B]$.

4. Bound (2.7) can be inverted to obtain bounds on quantiles of the distribution $P(y \mid x)$. Let α be any number between zero and one. Let $q(\alpha, x)$ denote the α-quantile of y conditional on x. Let $r(\alpha, x)$ denote the $[1 - (1 - \alpha)/P(z = 1 \mid x)]$-quantile of $P(y \mid x, z = 1)$ if $P(z = 1 \mid x) > 1 - \alpha$ and let $r(\alpha, x) = -\infty$ otherwise. Let $s(\alpha, x)$ denote the $[\alpha/P(z = 1 \mid x)]$-quantile of $P(y \mid x, z = 1)$ if $P(z = 1 \mid x) \geq \alpha$ and let $s(\alpha, x) = \infty$ otherwise. Then the result is this (see Manski, 1994a): $r(\alpha, x) \leq q(\alpha, x) \leq s(\alpha, x)$.

Observe that the lower and upper bounds are both increasing functions of the quantile α; hence the bound on $q(\alpha, x)$ shifts to the right as α increases. The lower bound is finite if the selection probability $P(z = 1 \mid x)$ is greater than $1 - \alpha$; the upper bound if $P(z = 1 \mid x) \geq \alpha$. So the bound restricts $q(\alpha, x)$ to an interval of finite width if $P(z = 1 \mid x)$ is greater than $\max(\alpha, 1 - \alpha)$. On the other hand, the bound is uninformative if the selection probability is less than $\min(\alpha, 1 - \alpha)$.

5. To obtain some intuition for this fact, consider the following thought experiment. Let t_0 and t_1 be real numbers, with $t_1 > t_0$. Let δ be a number between zero and one. Let w be a random variable with $P(w \leq t_0) = 1 - \delta$ and $P(w = t_1) = \delta$. Suppose that w is perturbed by moving the mass at t_1 to some $t_2 > t_1$. Then

$P(w \leq t)$ remains unchanged for $t < t_1$ and falls by at most δ for $t \geq t_1$. But $E(w)$ increases by the amount $\delta(t_2 - t_1)$. Now let t_2 increase to infinity. The perturbed distribution function remains within a δ-bound of the original one but the mean of the perturbed random variable increases to infinity.

6. I say "*seems* to turn downward" because this statement implicitly extrapolates the graph through regions of x where there are no observations.

7. This is the case if the conditioning variables x are not perfectly collinear and if the log-normality assumption is correct. If the log-normality assumption is incorrect, no value of the parameters generates the observable distributions.

8. It is particularly revealing to examine the bound (2.16) when selection is exogenous. (The researcher does not have this information but we can still examine its implications.) Then $P[y \in B \mid (w, v), z = 1] = P(y \in B \mid w)$ and (2.16) reduces to

$$(\star) \qquad P(y \in B \mid w)\pi \leq P(y \in B \mid w)$$
$$\leq P(y \in B \mid w)\pi + (1 - \pi),$$

where $\pi = \sup_v P(z = 1 \mid w, v)$. That is, π is the largest value taken by the selection probability, holding w fixed and letting v vary. The lower bound in $(\star)$ is strictly smaller than the upper bound whenever π is less than one. Thus, when selection is exogenous but the researcher does not know this, an exclusion restriction pins the value of $P(y \in B \mid w)$ down to a point only if there exists a value of v at which no censoring occurs.

9. Some authors have been concerned with inference on $P(y_1 - y_0 \mid x)$. For example, see Rubin (1974) and Rosenbaum (1984), who refer to $y_1 - y_0$ as the *causal effect* of treatment 1.

10. More generally, the widespread practice of interpreting regression contrasts as treatment effects rests on the assumption that selection into treatment is exogenous. Suppose that, given a random sample of observations of (y, x), one estimates $P(y \in B \mid x)$, computes $P(y \in B \mid x = x_1) - P(y \in B \mid x = x_0)$ for specified x_0 and x_1, and interprets this contrast as the change that would occur if a person with attributes x_0 were to be given attributes x_1 instead. This interpretation requires an exogenous-selection assumption.

To see why, let us recast the problem in the language of the treatment effects literature by assuming that each member of the population is characterized by values for the variables $[y(x), z(x), x \in X]$.

Here $z(x)$ is an indicator function taking the value 1 at a person's actual regressor value and 0 at all other of the logically possible regressor values X. Variable $y(x)$ is the outcome that would be observed if a person were to be assigned regressor value x. Of these outcomes, $y(x)$ is realized if and only if $z(x) = 1$. Thus the function $y(\cdot)$ is latent at all regressor values except the one that a person actually experiences. The realized outcome is

$$y = \sum_{x \in X} y(x) z(x).$$

This setup is the same as that of the classical treatment effect problem except that now there are more than two treatments; each value of x defines a different treatment.

Now define the treatment effect $P[y(x_1) \in B] - P[y(x_0) \in B]$ to be the change that would be observed if one were to replace a hypothetical situation in which a person were exogenously assigned regressor value x_0 with another hypothetical situation in which that person were exogenously assigned regressor value x_1. In general,

$$P(y \in B \mid x = x_1) - P(y \in B \mid x = x_0) = P[y(x_1) \in B \mid z(x_1) = 1]$$
$$- P[y(x_0) \in B \mid z(x_0) = 1]$$
$$\neq P[y(x_1) \in B]$$
$$- P[y(x_0) \in B].$$

The second inequality becomes an equality if the random outcome function $y(\cdot)$ is statistically independent of the random treatment-selection function $z(\cdot)$.

11. This finding is due to Heckman (1978). See also Heckman and Robb (1985) and Robinson (1989). To see why the result holds, decompose the mean of y_1 conditional on x into the sum

$$E(y_1 \mid x) = E(y_1 \mid x, z = 1) P(z = 1 \mid x)$$
$$+ E(y_1 \mid x, z = 0) P(z = 0 \mid x).$$

By (2.26), $E(y_1 \mid x, z = 0) = E(y_0 \mid x, z = 0) + k$. Hence

$$E(y_1 \mid x) = E(y_1 \mid x, z = 1) P(z = 1 \mid x)$$
$$+ E(y_0 \mid x, z = 0) P(z = 0 \mid x)$$
$$+ k P(z = 0 \mid x).$$

Let $x = (w, v)$ and, holding w fixed, impose the exclusion restriction $E(y_1 \mid w, v_0) = E(y_1 \mid w, v_1)$, where v_0 and v_1 are distinct values of v. Then

$$E[y_1 \mid (w, v_0), z = 1] P(z = 1 \mid w, v_0)$$
$$+ \ E[y_0 \mid (w, v_0), z = 0] P(z = 0 \mid w, v_0) + kP(z = 0 \mid w, v_0)$$
$$= E[y_1 \mid (w, v_1), z = 1] P(z = 1 \mid w, v_1)$$
$$+ \ E[y_0 \mid (w, v_1), z = 0] P(z = 0 \mid w, v_1) + kP(z = 0 \mid w, v_1).$$

This equation can be solved for k provided that $P(z = 0 \mid w, v_1) \neq P(z = 0 \mid w, v_0)$.

12. An economic application with the same structure is the short-side model of markets operating under regulated prices (see Maddala, 1983). Here y_1 is the quantity of a product that is demanded at the regulated price, and y_0 is the supply produced by firms. If demand is below supply, firms sell the quantity y_1 that consumers are willing to purchase and have some excess left over. If supply is below demand, firms sell all of their production y_0 and consumers are unable to make some desired purchases. So the quantity transacted is the minimum of y_1 and y_0. The inferential problem is to learn the distribution of demand $P(y_1 \mid x)$ and of supply $P(y_0 \mid x)$.

3. The Mixing Problem in Program Evaluation

1. From the mid-1960s through the late 1970s, social experiments were used to evaluate programs such as the negative income tax and time-of-day electricity pricing. Various experiments of this period are described in Hausman and Wise (1985). Welfare and training programs, however, were generally evaluated using observational data. One significant experimental evaluation was the National Supported Work Demonstration. See Manpower Demonstration Research Corporation (1980).
2. The analysis here is drawn from Manski (1994b).
3. One might observe realizations under more than one policy, of course. Work on selection problems has focused on the case in which only one policy is observed.
4. The mixing problem should not be confused with the converse problem: What does knowledge of $P[y_1 z_m + y_0(1 - z_m) \mid x]$ imply about $[P(y_1 \mid x), P(y_0 \mid x), P(z_m \mid x)]$? The latter is sometimes referred to as a *mixture* problem.

5. As in the selection problem, not all forms of prior information have identifying power. In the present situation, prior information restricting the marginal distributions $P(y_1 \mid x)$ and $P(y_0 \mid x)$ has no identifying power as these distributions are revealed by the empirical evidence. Such restrictions may improve the precision of sample estimates of $P(y_1 \mid x)$ and $P(y_0 \mid x)$, but this is not our present concern.

6. These estimates of $P(y_1 = 1 \mid x)$ and $P(y_0 = 1 \mid x)$ are based on the 58 treatment-group members and 63 control-group members from whom the investigators obtained graduation data.

7. See Ord (1972) for an exposition of the Frechet bound. It is elementary to show that $P(y_1 \in B \cap y_0 \in B \mid x)$ must satisfy the bound. The upper bound holds because the joint event $(y_1 \in B \cap y_0 \in B)$ implies each of its component events $(y_1 \in B)$ and $(y_0 \in B)$. The lower bound holds because

$$P(y_1 \in B \cup y_0 \in B \mid x) = P(y_1 \in B \mid x) + P(y_0 \in B \mid x)$$
$$- P(y_1 \in B \cap y_0 \in B \mid x) \leq 1.$$

The more subtle aspect of the Frechet bound is that it is sharp. That is, there is a joint distribution of (y_1, y_0) with marginals $P(y_1 \mid x)$ and $P(y_0 \mid x)$ for which $P(y_1 \in B \cap y_0 \in B \mid x)$ equals the lower bound in (3.8), and there is another distribution with these marginals for which $P(y_1 \in B \cap y_0 \in B \mid x)$ equals the upper bound in (3.8).

8. Knowledge of $P(y_1 \mid x)$ and $P(y_0 \mid x)$ makes the shifted-outcome assumption a testable hypothesis. If (3.13) holds, $P(y_1 \mid x)$ and $P(y_0 \mid x)$ must be the same up to a translation of location. In contrast, the assumption that outcomes are statistically independent is not testable, as it implies no restrictions on $P(y_1 \mid x)$ and $P(y_0 \mid x)$.

9. Knowledge of $P(y_1 \mid x)$ and $P(y_0 \mid x)$ makes the ordered-outcomes assumption a testable hypothesis. If (3.16) holds, then $P(y_0 \in B \mid x) \geq P(y_1 \in B \mid x)$.

4. Response-Based Sampling

1. This negative result holds whenever $P(w, r \mid y = 1) > 0$ and $P(w, r \mid y = 0) > 0$, but does not hold if one of these probabilities equals zero. If $P(w, r \mid y = 1) > 0$ but $P(w, r \mid y = 0) = 0$, then (4.3) implies that $P(y = 1 \mid w, r) = 1$. If $P(w, r \mid y = 1) = 0$ but

$P(w, r \mid y = 0) > 0$, then $P(y = 1 \mid w, r) = 0$. These positive identification findings are degenerate exceptions to the general rule that response-based sampling data do not identify response probabilities.

2. To prove (4.11), we show that relative risk is monotone in $P(y = 1 \mid w)$, the direction of change depending on the magnitude of the odds ratio. To see this, let $p \equiv P(y = 1 \mid w)$ and let $P_{im} \equiv P(r = i \mid w, y = m)$ for $i = j$, k, and $m = 0, 1$. Write the relative risk (4.5) explicitly as a function of p. Thus, define

$$RR_p = \frac{P_{k1}}{P_{j1}} \frac{(P_{j1} - P_{j0})p + P_{j0}}{(P_{k1} - P_{k0})p + P_{k0}} .$$

The derivative of RR_p with respect to p is

$$\frac{P_{k1}}{P_{j1}} \frac{P_{j1}P_{k0} - P_{k1}P_{j0}}{[(P_{k1} - P_{k0})p + P_{k0}]^2} .$$

This derivative is positive if OR $<$ 1, zero if OR $=$ 1, and negative if OR $>$ 1.

The fact that relative risk is monotone in $P(y = 1 \mid w)$ implies that the extreme values occur when $P(y = 1 \mid w)$ equals its extreme values of 0 and 1. Setting $P(y = 1 \mid w) = 0$ makes RR $=$ OR, and setting $P(y = 1 \mid w) = 1$ makes RR $=$ 1.

3. This follows from the fact, shown in note 2, that RR_p is monotone in p.

4. To prove (4.15), we show that as $P(y = 1 \mid w)$ increases from 0 to 1, the attributable risk changes parabolically, the orientation of the parabola depending on the magnitude of the odds ratio. To see this, again let $p \equiv P(y = 1 \mid w)$ and $P_{im} \equiv P(r = i \mid w, y = m)$ for $i = j$, k, and $m = 0, 1$. Write the attributable risk explicitly as a function of p, as in (4.12). Thus let

$$AR_p \equiv \frac{P_{k1}p}{(P_{k1} - P_{k0})p + P_{k0}} - \frac{P_{j1}p}{(P_{j1} - P_{j0})p + P_{j0}} .$$

The derivative of AR_p with respect to p is

$$\frac{P_{k1}P_{k0}}{[(P_{k1} - P_{k0})p + P_{k0}]^2} - \frac{P_{j1}P_{j0}}{[(P_{j1} - P_{j0})p + P_{j0}]^2} .$$

The derivative equals zero at

$$\pi = \frac{\beta P_{k0} - P_{j0}}{(\beta P_{k0} - P_{j0}) - (\beta P_{k1} - P_{j1})}$$

and at

$$\pi^\star = \frac{\beta P_{k0} + P_{j0}}{(\beta P_{k0} + P_{j0}) - (\beta P_{k1} + P_{j1})},$$

where $\beta \equiv (P_{j1} P_{j0}/ P_{k1} P_{k0})^{1/2}$ as in (4.14). Examination of the two roots reveals that π always lies between zero and one, but $\pi^\star$ always lies outside the unit interval; so π is the only relevant root.

Thus AR_p varies parabolically as p rises from zero to one. Observe that $AR_p = 0$ at $p = 0$ and at $p = 1$. Examination of the derivative of AR_p at $p = 0$ and at $p = 1$ shows that the orientation of the parabola depends on the magnitude of the odds ratio. If OR is less than one, then as p rises from zero to one, AR_p falls from zero to its minimum at π and then rises back to zero. If the odds ratio is greater than one, then AR_p rises from zero to its maximum at π and then falls back to zero. In the borderline case where the odds ratio equals one, AR_p does not vary with p.

5. Predicting Individual Behavior

1. If β equals zero, then students are always indifferent between schooling and work, whatever values (A, I, Q, C, W) may take. This is implausible. Note that if β really were to equal zero, then the rational-choice model would imply no restrictions on observed behavior.

2. The assumption that u is continuously distributed simplifies revealed preference analysis and is imposed routinely in empirical studies. The random utility model makes no prediction of behavior when a person is indifferent between the available alternatives; that is, when $U_1 = U_0$. If u is continuously distributed, then $P(U_1 = U_0 \mid x) = 0$. Hence the researcher need not be concerned with how persons behave when they are indifferent.

3. Random utility models were originally developed by the psychologist Thurstone (1927), who did think of behavior as inherently

probabilistic. Luce and Suppes (1965) survey the subsequent psychological literature. The econometric derivation given here is due to McFadden (1973).

4. Normalizations of location and scale are innocuous as distributions differing only in location and scale are indistinguishable from one another. To see this, let $v = au + b$, where a is any positive number and b is any real number. The condition $x\beta + u \geq 0$ holds if and only if $x(a\beta) - b + v \geq 0$. Thus the random utility model with u as the unobserved variable is indistinguishable from one with v as the unobserved variable.

5. It is important to determine what extrapolations may be made when different distributional assumptions are imposed on u. This question has drawn considerable theoretical attention, and much has been learned. I describe some basic ideas here, drawing on Manski (1988b).

Random-sample data on (y, x) identify $P(y \mid x)$ on the support of $P(x)$. At each value of x, whether on the support or off, the representation of $P(y \mid x)$ on the right side of (5.6) is a function of β and of the distribution $P(u \mid x)$. So assertion of (5.6) raises this inferential question: What does knowledge of $P(y \mid x)$ imply about β and $P(u \mid x)$? The literature on identification of binary response models analyzes this question and translates findings on identification of β and $P(u \mid x)$ into findings on extrapolation.

Consider first the prevailing empirical practice, wherein u is assumed to be statistically independent of x with a distribution known up to normalizations of location and scale. With $P(u \mid x)$ known, the only inferential problem is to learn β. Given random-sample data on (y, x), the left side of (5.6) is identified on the support of $P(x)$. The parameters β are identified if there exists exactly one non-zero value of β that satisfies equation (5.6) at all x on the support. The random utility model is misspecified if there exists no value of β solving (5.6) and is not identified if there exist multiple values. It can be shown that β is identified if and only if the model is correctly specified and the support of $P(x)$ contains at least K linearly independent values, where K is the number of components in x. When these conditions are satisfied, equation (5.6) can be used to determine $P(y = 1 \mid x)$ at any value of x, whether on or off the support.

Now drop the conventional distributional assumption and assume only that u is *median independent* of x. Set the median of u equal to zero to normalize location. Then $P(u \leq 0 \mid x) = P(u \geq 0 \mid x) = 1/2$ for all values of x. It follows that

$$(\star) \qquad P(y = 1 \mid x) < 1/2 \Rightarrow P(x\beta + u \geq 0 \mid x) < 1/2$$

$$\Rightarrow x\beta \leq 0$$

$$P(y = 1 \mid x) > 1/2 \Rightarrow P(x\beta + u \geq 0 \mid x) > 1/2$$

$$\Rightarrow x\beta \geq 0.$$

Random sampling reveals whether $P(y = 1 \mid x)$ lies below or above $1/2$ at each value of x on the support. So the possible values of β are those that make $(\star)$ hold on the support. Assuming that the model is correctly specified, β is identified up to scale if the support of $P(x)$ has at least K linearly independent points and if there exists at least one component of x, say, x_K, for which $\beta_K \neq 0$ and the support of $P(x_K)$ is the whole real line. With β identified up to scale, the sign of $x\beta$ may be used to determine whether $P(y = 1 \mid x)$ is below or above $1/2$ at any value of x, whether on or off the support. But the actual value of $P(y = 1 \mid x)$ is not identified. There is thus a clear tradeoff between assuming that u is statistically independent of x with known distribution and assuming only that u is median independent. The former assumption is stronger but yields more extrapolation power.

6. This discussion is drawn from Manski (1993b).

7. In the final chapters of his book, Freeman reported findings from a one-time survey of college students regarding their income expectations in various occupations. But his analysis of these data sheds no light on the realism of the myopic-expectations assumption made earlier on.

8. The idea that providing information can influence schooling behavior is reflected in traditional school counseling programs and in the recent spread of mentoring programs. Moreover, it underlies the widespread belief that youth are influenced by the "role models" in their environment. It is often argued that if youth learn about the returns to schooling from observing the experiences of their peers, families, and neighbors, then schooling behavior can be influenced by altering the role models that youth observe.

9. For example, Ogbu (1978) states that "blacks, from generations of experience, realize that they face a job ceiling." Rosenbaum and Kariya (1989) assert that "since grades have little influence on youth's wages or jobs . . . , school performance has little payoff in any kind of job attainments."

10. The idea of using incentives to influence schooling behavior is reflected in traditional policies ranging from the honor roll through corporal punishment. Recently enacted programs bring to bear positive and negative financial incentives. On the one hand, programs such as those promoted by the "I Have a Dream" Foundation guarantee students college scholarships if they perform well in high school (Lacey, 1989). On the other hand, Wisconsin's "Learnfare" policy sanctions families receiving AFDC payments if their children drop out of high school or do not maintain normal attendance levels (Corbett et al., 1989).

11. The U.S. General Accounting Office (1990), in an overview of private programs offering information and incentives to youth at risk, observes that data permitting a systematic evaluation of these programs are not available.

12. Interestingly, economists do not apply this reasoning to self-reports of "objective" data. Empirical economic analyses routinely accept as fact respondents' reports of their socioeconomic/demographic characteristics, choices, and outcomes. Thus economists' own revealed preferences in empirical analysis are somewhat inconsistent with their expressed views about the interpretability of survey data.

13. Intentions questions are sometimes confused with *forced-choice* questions. The latter type of question requires the respondent to state what choice he would make if required to commit himself to a decision at the time of the survey. Voter election surveys illustrate the distinction well. Some surveys ask:

> "For whom do you expect to vote in the coming election, candidate 0 or candidate 1?"

Others ask:

> "For whom would you vote if you had to cast your ballot today, candidate 0 or candidate 1?"

The former is an intentions question, the latter a forced-choice one. A person with rational expectations need not give the same response to these two questions. The intentions response is the person's prediction of the decision he will make when he has the information s and u. The forced-choice response is the person's decision given only the information s available at the time of the survey. See Manski (1990b) for further discussion.

14. One difference is that a "temporary lay-off" question was added to the 1974 survey. A second is that the instructions call for respondents

to "circle as many as apply" rather than to "circle one number on each line." A third difference is that the respondents to the intentions questions were asked to predict their behavior during October 1974, whereas the behavior questions concern only the first week of that month. These distinctions will be ignored here, although it is possible that they are germane.

6. Simultaneity

1. The identification argument given here is the standard one found in econometrics textbooks. Actually, this argument imposes much stronger assumptions than are necessary. Consider the problem of inference on demand. Assume only that demand is a linear function of price, with the same slope β_2 in each market and an intercept ε_2 that may vary across markets; thus

   (⋆) $d(t) = \beta_2 t + \varepsilon_2.$

   Impose an exclusion restriction of the form

   (⋆⋆) $E(\varepsilon_2 \mid x = x_i) = E(\varepsilon_2 \mid x = x_j)$

   (⋆⋆⋆) $E(p \mid x = x_i) \neq E(p \mid x = x_j),$

   where x_i and x_j are known points on the support of the distribution of covariates. Conditions (6.1), (⋆), (⋆⋆), and (⋆⋆⋆) suffice to identify $P[d(\cdot) \mid x]$.

   To see this, use (6.1) and (⋆) to write

   $$\varepsilon_2 = q - \beta_2 p.$$

   Inserting this expression for ε_2 into (⋆⋆) yields

   $$E(q - \beta_2 p \mid x = x_i) = E(q - \beta_2 p \mid x = x_j).$$

   Provided that (⋆⋆⋆) holds, this equation can be solved for β_2. The result is

   $$\beta_2 = \frac{E(q \mid x = x_i) - E(q \mid x = x_j)}{E(p \mid x = x_i) - E(p \mid x = x_j)}.$$

   The sampling process identifies the right-side expression, hence β_2.

   Knowing β_2 implies that ε_2 is identified in each market whose outcome (p, q) is observed. Knowing β_2 and ε_2 implies knowledge of the demand function $d(\cdot)$. Knowing $[d(\cdot), x]$ in a random sample of markets reveals $P[d(\cdot) \mid x]$.

2. Here I summarize Manski (1994c).

3. Equation (6.21) is identical to equation (2.28) except that the assumed direction of the inequality is reversed here.

4. Section 2.6 also considered the *shifted-outcomes* assumption; that is, $y_1 = y_0 + \beta$, where β is a parameter. This is the simplest nontrivial case of the linear demand assumption (6.3b), obtained by evaluating (6.3b) at $t = 0$ and at $t = 1$.

5. A complementary result holds for upward-sloping functions. Assume $d(\cdot)$ to be upward sloping rather than downward sloping. Then observations in the SE and NW regions are informative, but those in the SW and NE regions are not.

6. The *County and City Data Book 1988* lists twenty-three Wisconsin cities as having populations over 25,000, but statistics are not reported for one of these, the city of Superior.

7. Specification of the covariates x should reflect one's beliefs about the variation in crime functions across cities. For example, suppose one believes that the distribution of crime functions across Wisconsin cities is the same as the distribution of crime functions across all midwestern cities. Then one can let $x =$ (midwestern cities) and use the Wisconsin data to infer the distribution of crime functions across all midwestern cities.

7. The Reflection Problem

1. Although it is valid to distinguish mainstream economic thinking on social effects from the perspectives of the other social sciences, one should not think that economists are concerned only with the operation of markets. The field of public economics has long been concerned with *external effects,* or the social effects on opportunities that operate outside markets. Moreover, some economists have sought to interpret and make use of key sociological ideas. Duesenberry (1949) is one example. More recently, Schelling (1971) analyzed the residential patterns that emerge when individuals choose not to live in neighborhoods where the percentage of residents of their own race is below some threshold. Conlisk (1980) showed that, if decision making is costly, it may be optimal for individuals to imitate the behavior of other persons who are better informed. Akerlof (1980) and Jones (1984) studied the equilibria of noncooperative games in which individuals are punished for deviation from group norms. Gaertner (1974), Pollak (1976), Alessie and Kapteyn (1991), and Case (1991) analyzed consumer demand models in

which, holding price fixed, individual demand increases with the mean demand of a reference group.

2. This chapter draws on Manski (1993c).

3. Beginning with Hyman (1942), reference-group theory has sought to express the idea that individuals learn from or are otherwise influenced by the behavior and attitudes of some reference group. Bank, Slavings, and Biddle (1990) give a historical account. Sociological writing has remained verbal, but economists have interpreted reference groups as conditioning variables, in the manner of (7.2). See Alessie and Kapteyn (1991) or Manski (1993d).

4. The regressors x are discrete in many empirical studies. In others, it is appropriate to assume that the regressions are continuous in x. Nevertheless, there are cases of empirical interest where neither condition holds. In particular, they break down in studies of small-group social interactions, such as family interactions. Here each reference group (that is, family) has negligible size relative to the population, and random sampling of individuals only rarely yields multiple members of the same family. So it is not a good empirical approximation to assume that x is discrete. Moreover, unless one can characterize groups of families as being similar in composition, it is not plausible to assume that $E(y \mid x)$ and $E(z \mid x)$ are continuous functions of x. The conclusion to be drawn, not surprisingly, is that random sampling of individuals is not an effective data-gathering process for the study of family interactions. It is preferable to use families as the sampling unit. The properties of alternative sample designs for the study of small-group interactions have been examined in the literature on network sampling. See Marsden (1990) for a review article.

References

Ajzen, I., and M. Fishbein. 1980. *Understanding Attitudes and Predicting Social Behavior.* Englewood Cliffs, N.J.: Prentice-Hall.

Akerlof, G. 1980. "A Theory of Social Custom, of Which Unemployment May Be One Consequence." *Quarterly Journal of Economics,* 94: 749–775.

Alessie, R., and A. Kapteyn. 1991. "Habit Formation, Interdependent Preferences, and Demographic Effects in the Almost Ideal Demand System." *Economic Journal,* 101: 404–419.

Arabmazar, A., and P. Schmidt. 1982. "An Investigation of the Robustness of the Tobit Estimator to Non-normality." *Econometrica,* 50: 1055–1063.

Bank, B., R. Slavings, and B. Biddle. 1990. "Effects of Peer, Faculty, and Parental Influences on Students' Persistence." *Sociology of Education,* 63: 208–225.

Bassi, L., and O. Ashenfelter. 1986. "The Effect of Direct Job Creation and Training Programs on Low-Skilled Workers." In S. Danziger and D. Weinberg, eds., *Fighting Poverty.* Cambridge, Mass.: Harvard University Press.

Berkson, J. 1958. "Smoking and Lung Cancer: Some Observations on Two Recent Reports." *Journal of the American Statistical Association,* 53: 28–38.

Berrueta-Clement, J., L. Schweinhart, W. Barnett, A. Epstein, and D. Weikart. 1984. *Changed Lives: The Effects of the Perry Preschool Program on Youths through Age 19.* Ypsilanti, Mich.: High/Scope Press.

Blackmore, J., and J. Welsh. 1983. "Selective Incapacitation: Sentencing According to Risk." *Crime and Delinquency,* 29: 504–528.

Blumstein, A., J. Cohen, and D. Nagin, eds. 1978. *Deterrence and Incapacitation: Estimating the Effects of Criminal Sanctions on Crime Rates.* Washington, D.C.: National Academy Press.

Blumstein, A., J. Cohen, J. Roth, and C. Visher, eds. 1986. *Criminal Careers and Career Criminals.* Washington, D.C.: National Academy Press.

Borjas, G. 1992. "Ethnic Capital and Intergenerational Mobility." *Quarterly Journal of Economics,* 107: 123–150.

Bork, R. (Solicitor General) et al. 1974. *Fowler v. North Carolina,* U.S. Supreme Court case no. 73-7031. Brief for U.S. as *amicus curiae:* 32–39.

Bowers, W., and G. Pierce. 1975. "The Illusion of Deterrence in Isaac Ehrlich's Research on Capital Punishment." *Yale Law Journal,* 85: 187–208.

Campbell, D., and J. Stanley. 1963. *Experimental and Quasi-Experimental Designs.* Chicago: Rand McNally.

Case, A. 1991. "Spatial Patterns in Household Demand." *Econometrica,* 59: 953–965.

Chaiken, J., and M. Chaiken. 1982. *Varieties of Criminal Behavior.* Report R-2814-NIJ. Santa Monica, Calif.: Rand Corporation.

Cliff, A., and J. Ord. 1981. *Spatial Processes.* London: Pion.

Cochran, W. 1977. *Sampling Techniques.* 3rd ed. New York: Wiley.

Cochran, W., F. Mosteller, and J. Tukey. 1954. *Statistical Problems of the Kinsey Report on Sexual Behavior in the Human Male.* Washington, D.C.: American Statistical Association.

Coleman, J., E. Campbell, C. Hobson, J. McPartland, A. Mood, F. Weinfeld, and R. York. 1966. *Equality of Educational Opportunity.* Washington, D.C.: U.S. Government Printing Office.

Coleman, J., E. Katz, and H. Menzel. 1957. "The Diffusion of an Innovation among Physicians." *Sociometry,* 20: 253–270.

Conlisk, J. 1980. "Costly Optimizers versus Cheap Imitators." *Journal of Economic Behavior and Organization,* 1: 275–293.

Corbett, T., J. Deloya, W. Manning, and L. Uhr. 1989. "Learnfare: The Wisconsin Experience." *Focus,* 12(2): 1–10.

Cornfield, J. 1951. "A Method of Estimating Comparative Rates from Clinical Data: Applications to Cancer of the Lung, Breast, and Cervix." *Journal of the National Cancer Institute,* 11: 1269–1275.

Coyle, S., R. Boruch, and C. Turner, eds. 1989. *Evaluating AIDS Prevention Programs.* Washington, D.C.: National Academy Press.

Crane, J. 1991. "The Epidemic Theory of Ghettos and Neighborhood

Effects on Dropping Out and Teenage Childbearing." *American Journal of Sociology*, 96: 1226–1259.

Davidson, A., and L. Beach. 1981. "Error Patterns in the Prediction of Fertility Behavior." *Journal of Applied Social Psychology*, 11: 475–488.

Davidson, A., and J. Jaccard. 1979. "Variables That Moderate the Attitude-Behavior Relation: Results of a Longitudinal Survey." *Journal of Personality and Social Psychology*, 37: 1364–1376.

Doreian, Patrick. 1981. "Estimating Linear Models with Spatially Distributed Data." *Sociological Methodology 1981*, 11: 359–388.

Duesenberry, J. 1949. *Income, Savings, and the Theory of Consumption*. Cambridge, Mass.: Harvard University Press.

Duncan, O., A. Haller, and A. Portes. 1968. "Peer Influences on Aspirations: A Reinterpretation." *American Journal of Sociology*, 74: 119–137.

Duncan, O., A. Haller, and A. Portes. 1970. "Duncan's Corrections of the Published Text to 'Peer Influences on Aspirations: A Reinterpretation.' " *American Journal of Sociology*, 75: 1042–1046.

Ehrlich, I. 1975. "The Deterrent Effect of Capital Punishment: A Question of Life and Death." *American Economic Review*, 65: 397–417.

Erbring, L., and A. Young. 1979. "Individuals and Social Structure: Contextual Effects as Endogenous Feedback." *Sociological Methods and Research*, 7: 396–430.

Ferguson, T. 1967. *Mathematical Statistics: A Decision-Theoretic Approach*. New York: Academic Press.

Fishbein, M., and I. Ajzen. 1975. *Belief, Attitude, Intention, and Behavior: An Introduction to Theory and Research*. Reading, Mass.: Addison-Wesley.

Fisher, F. 1966. *The Identification Problem in Econometrics*. New York: McGraw-Hill.

Fisher, R. 1935. *The Design of Experiments*. London: Oliver and Boyd.

Fleiss, J. 1981. *Statistical Methods for Rates and Proportions*. New York: Wiley.

Frechet, M. 1951. "Sur les Tableaux de Correlation Donte les Marges Sont Données." *Annals de Université de Lyon A*, ser. 3, 14: 53–77.

Freeman, R. 1971. *The Market for College-Trained Manpower*. Cambridge, Mass.: Harvard University Press.

Friedkin, N. 1990. "Social Networks in Structural Equation Models." *Social Psychology Quarterly*, 53: 316–328.

Friedman, M. 1953. *Essays in Positive Economics*. Chicago: University of Chicago Press.

Friedman, M. 1957. *A Theory of the Consumption Function*. Princeton: Princeton University Press.

Gaertner, W. 1974. "A Dynamic Model of Interdependent Consumer Behavior." *Zeitschrift für Nationalokonomie*, 70: 312–326.

Gamoran, A. 1992. "Social Factors in Education." In M. Alkin, ed., *Encyclopedia of Educational Research*. 6th ed. New York: Macmillan.

Garfinkel, I., C. Manski, and C. Michalopolous. 1992. "Micro-Experiments and Macro Effects." In C. Manski and I. Garfinkel, eds., *Evaluating Welfare and Training Programs*. Cambridge, Mass.: Harvard University Press.

Goldberger, A. 1983. "Abnormal Selection Bias." In S. Karlin, T. Amemiya, and L. Goodman, eds., *Studies in Econometrics, Time Series, and Multivariate Statistics*. New York: Academic Press.

Greenberg, D., and M. Wiseman. 1992. "What Did the OBRA Demonstrations Do?" In C. Manski and I. Garfinkel, eds., *Evaluating Welfare and Training Programs*. Cambridge, Mass.: Harvard University Press.

Greenwood, P. 1982. *Selective Incapacitation*. Report R-2815-NIJ. Santa Monica, Calif.: Rand Corporation.

Gronau, R. 1974. "Wage Comparisons—a Selectivity Bias." *Journal of Political Economy*, 82: 1119–1143.

Hanushek, E. 1986. "The Economics of Schooling." *Journal of Economic Literature*, 24: 1141–1177.

Hardle, W. 1990. *Applied Nonparametric Regression*. Cambridge: Cambridge University Press.

Hauser, R. 1970. "Context and Consex: A Cautionary Tale." *American Journal of Sociology*, 75: 645–664.

Hausman, J., and D. Wise, eds. 1985. *Social Experimentation*. Chicago: University of Chicago Press.

Hayes, C., and S. Hofferth, eds. 1987. *Risking the Future: Adolescent Sexuality, Pregnancy, and Childbearing*. Washington, D.C.: National Academy Press.

Heckman, J. 1976. "The Common Structure of Statistical Models of Truncation, Sample Selection, and Limited Dependent Variables and a Simple Estimator for Such Models." *Annals of Economic and Social Measurement*, 5: 479–492.

Heckman, J. 1978. "Dummy Endogenous Variables in a Simultaneous Equation System." *Econometrica*, 46: 931–959.

Heckman, J. 1992. "Randomization and Social Policy Evaluation." In

C. Manski and I. Garfinkel, eds., *Evaluating Welfare and Training Programs*. Cambridge, Mass.: Harvard University Press.

Heckman, J., and R. Robb. 1985. "Alternative Methods for Evaluating the Impact of Interventions." In J. Heckman and B. Singer, eds., *Longitudinal Analysis of Labor Market Data*. Cambridge: Cambridge University Press.

Hendershot, G., and P. Placek, eds. 1981. *Predicting Fertility*. Lexington, Mass.: D. C. Heath.

Holden, C. 1990. "Head Start Enters Adulthood." *Science*, 247: 1400–1402.

Hood, W., and T. Koopmans, eds. 1953. *Studies in Econometric Method*. New York: Wiley.

Hotz, J. 1992. "Designing an Evaluation of the Job Training Partnership Act." In C. Manski and I. Garfinkel, eds., *Evaluating Welfare and Training Programs*. Cambridge, Mass.: Harvard University Press.

Hsieh, D., C. Manski, and D. McFadden. 1985. "Estimation of Response Probabilities from Augmented Retrospective Observations." *Journal of the American Statistical Association*, 80: 651–662.

Huber, P. 1981. *Robust Statistics*. New York: Wiley.

Hurd, M. 1979. "Estimation in Truncated Samples When There Is Heteroskedasticity." *Journal of Econometrics*, 11: 247–258.

Hyman, H. 1942. "The Psychology of Status." *Archives of Psychology*, no. 269.

Jamieson, L., and F. Bass. 1989. "Adjusting Stated Intentions Measures to Predict Trial Purchase of New Products: A Comparison of Models and Methods." *Journal of Marketing Research*, 26: 336–345.

Jencks, C., and S. Mayer. 1989. "Growing Up in Poor Neighborhoods: How Much Does It Matter?" *Science*, 243: 1441–1445.

Jones, S. 1984. *The Economics of Conformism*. Oxford: Basil Blackwell.

Juster, T. 1964. *Anticipations and Purchases*. Princeton: Princeton University Press.

Juster, T. 1966. "Consumer Buying Intentions and Purchase Probability: An Experiment in Survey Design." *Journal of the American Statistical Association*, 61: 658–696.

Kalbfleisch, J., and R. Prentice. 1980. *The Statistical Analysis of Failure Time Data*. New York: Wiley.

Klein, L., B. Forst, and V. Filatov. 1978. "The Deterrent Effect of Capital Punishment: An Assessment of the Estimates." In A. Blumstein, J. Cohen, and D. Nagin, eds., *Deterrence and Incapacitation: Estimating the Effects of Criminal Sanctions on Crime Rates*. Washington, D.C.: National Academy Press.

Koopmans, T. 1949. "Identification Problems in Economic Model Construction." *Econometrica*, 17: 125–144.

Lacey, R. 1989. "From Inspiration to Institution: The 'I Have a Dream' Approach to Dropout Prevention." *National Center on Effective Secondary Schools Newsletter*, 4(3): 3–8.

LaLonde, R. 1986. "Evaluating the Econometric Evaluations of Training Programs with Experimental Data." *American Economic Review*, 76: 604–620.

Luce, R. D., and P. Suppes. 1965. "Preference, Utility, and Subjective Probability." In R. D. Luce, R. Bush, and E. Galanter, eds., *Handbook of Mathematical Psychology*. Vol. 3. New York: Wiley.

Maddala, G. S. 1983. *Qualitative and Limited Dependent Variable Models in Econometrics*. Cambridge: Cambridge University Press.

Manpower Demonstration Research Corporation. 1980. *Summary and Findings of the National Supported Work Demonstration*. Cambridge, Mass.: Ballinger.

Manski, C. 1988a. *Analog Estimation Methods in Econometrics*. London: Chapman and Hall.

Manski, C. 1988b. "Identification of Binary Response Models." *Journal of the American Statistical Association*, 83: 729–738.

Manski, C. 1989. "Anatomy of the Selection Problem." *Journal of Human Resources*, 24: 343–360.

Manski, C. 1990a. "Nonparametric Bounds on Treatment Effects." *American Economic Review Papers and Proceedings*, 80: 319–323.

Manski, C. 1990b. "The Use of Intentions Data to Predict Behavior: A Best Case Analysis." *Journal of the American Statistical Association*, 85: 934–940.

Manski, C. 1991. "Regression." *Journal of Economic Literature*, 29: 34–50.

Manski, C. 1993a. "Identification Problems in the Social Sciences." *Sociological Methodology*, 23: 1–56.

Manski, C. 1993b. "Adolescent Econometricians: How Do Youth Infer the Returns to Schooling?" In C. Clotfelter and M. Rothschild, eds., *Studies of Supply and Demand in Higher Education*. Chicago: University of Chicago Press.

Manski, C. 1993c. "Identification of Endogenous Social Effects: The Reflection Problem." *Review of Economic Studies*, 60: 531–542.

Manski, C. 1993d. "Dynamic Choice in a Social Setting: Learning from the Experiences of Others." *Journal of Econometrics*, 58:121–136.

Manski, C. 1994a. "The Selection Problem." In C. Sims, ed., *Advances in Econometrics*. Cambridge: Cambridge University Press.

Manski, C. 1994b. "The Mixing Problem in Program Evaluation." Social Systems Research Institute Working Paper 9313R, University of Wisconsin–Madison.

Manski, C. 1994c. "Simultaneity with Downward Sloping Demand." Social Systems Research Institute Working Paper 9408, University of Wisconsin–Madison.

Manski, C., and I. Garfinkel, eds. 1992. *Evaluating Welfare and Training Programs.* Cambridge, Mass.: Harvard University Press.

Manski, C., and S. Lerman. 1977. "The Estimation of Choice Probabilities from Choice-Based Samples." *Econometrica,* 45: 1977–1988.

Manski, C., G. Sandefur, S. McLanahan, and D. Powers. 1992. "Alternative Estimates of the Effect of Family Structure during Adolescence on High School Graduation." *Journal of the American Statistical Association,* 87: 25–37.

Manski, C., and D. Wise. 1983. *College Choice in America.* Cambridge, Mass.: Harvard University Press.

Marschak, J. 1953. "Economic Measurements for Policy and Prediction." In W. Hood and T. Koopmans, eds., *Studies in Econometric Method.* New York: Wiley.

Marsden, P. 1990. "Network Data and Measurement." *Annual Review of Sociology,* 16: 435–463.

Mayer, S. 1991. "How Much Does a High School's Racial and Socioeconomic Mix Affect Graduation and Teenage Fertility Rates?" In C. Jencks and P. Peterson, eds., *The Urban Underclass.* Washington, D.C.: Brookings Institution.

McFadden, D. 1973. "Conditional Logit Analysis of Qualitative Choice Behavior." In P. Zarembka, ed., *Advances in Econometrics.* New York: Academic Press.

McLanahan, S., and G. Sandefur. 1994. *Growing Up with a Single Parent.* Cambridge, Mass.: Harvard University Press.

Moffitt, R. 1992a. "Incentive Effects of the U.S. Welfare System: A Review." *Journal of Economic Literature,* 30: 1–61.

Moffitt, R. 1992b. "Evaluation Methods for Program Entry Effects." In C. Manski and I. Garfinkel, eds., *Evaluating Welfare and Training Programs.* Cambridge, Mass.: Harvard University Press.

Morrison, D. 1979. "Purchase Intentions and Purchase Behavior." *Journal of Marketing,* 43: 65–74.

Murphy, K., and F. Welch. 1989. "Wage Premiums for College Graduates: Recent Growth and Possible Explanations." *Educational Researcher,* 18(4): 17–26.

1993 *Green Book. See* U.S. House of Representatives, 1993.

O'Connell, M., and C. Rogers. 1983. "Assessing Cohort Birth Expectations Data from the Current Population Survey, 1971–1981." *Demography,* 20: 369–383.

Ogbu, J. 1978. *Minority Education and Caste: The American System in Cross-Cultural Perspective.* New York: Academic Press.

Ord, J. 1972. *Families of Frequency Distributions.* Griffin's Statistical Monographs and Courses no. 30. New York: Hafner.

Passell, P., and J. Taylor. 1975. "The Deterrent Effect of Capital Punishment: Another View." Discussion Paper 74-7509, Department of Economics, Columbia University, New York.

Piliavin, I., and M. Sosin. 1988. "Exiting Homelessness: Some Recent Empirical Findings." Institute for Research on Poverty, University of Wisconsin–Madison.

Pollak, R. 1976. "Interdependent Preferences." *American Economic Review,* 78: 745–763.

Riccobono, J., L. Henderson, G. Burkheimer, C. Place, and J. Levinsohn. 1981. *National Longitudinal Study: Data File Users Manual.* Washington, D.C.: National Center for Education Statistics, U.S. Department of Education.

Robinson, C. 1989. "The Joint Determination of Union Status and Union Wage Effects: Some Tests of Alternative Models." *Journal of Political Economy,* 97: 639–667.

Rosenbaum, J., and T. Kariya. 1989. "From High School to Work: Market and Institutional Mechanisms in Japan." *American Journal of Sociology,* 94: 1334–1365.

Rosenbaum, P. 1984. "From Association to Causation in Observational Studies: The Role of Tests of Strongly Ignorable Treatment Assignment." *Journal of the American Statistical Association,* 79: 41–48.

Rosenbaum, P., and D. Rubin. 1983. "The Central Role of the Propensity Score in Observational Studies for Causal Effects." *Biometrika,* 70: 41–55.

Roy, A. 1951. "Some Thoughts on the Distribution of Earnings." *Oxford Economic Papers,* 3: 135–146.

Rubin, D. 1974. "Estimating Causal Effects of Treatments in Randomized and Nonrandomized Studies." *Journal of Educational Psychology,* 66: 688–701.

Schelling, T. 1971. "Dynamic Models of Segregation." *Journal of Mathematical Sociology,* 1: 143–186.

Schuman, H., and M. Johnson. 1976. "Attitudes and Behavior." *Annual Review of Sociology,* 2: 161–207.

Sewell, W., and J. Armer. 1966. "Neighborhood Context and College Plans." *American Sociological Review,* 31: 159–168.

Stigler, S. 1986. *The History of Statistics.* Cambridge, Mass.: Harvard University Press.

Thurstone, L. 1927. "A Law of Comparative Judgment." *Psychological Review,* 34: 273–286.

Turner, C., and E. Martin, eds. 1984. *Surveying Subjective Phenomena.* New York: Russell Sage Foundation.

Urban, G., and J. Hauser. 1980. *Design and Marketing of New Products.* Englewood Cliffs, N.J.: Prentice-Hall.

U.S. Bureau of the Census. 1988a. *Fertility of American Women: June, 1987.* Current Population Reports, series P-20, no. 427. Washington, D.C.: U.S. Government Printing Office.

U.S. Bureau of the Census. 1988b. *County and City Data Book 1988.* Washington, D.C.: U.S. Government Printing Office.

U.S. Bureau of the Census. 1991. *Money Income of Households, Families, and Persons in the United States: 1988 and 1989.* Current Population Reports, series P-60, no. 172. Washington, D.C.: U.S. Government Printing Office.

U.S. General Accounting Office. 1990. *Promising Practice: Private Programs Guaranteeing Student Aid for Higher Education.* GAO/PEMD-90-16. Gaithersburg, Md.: U.S. General Accounting Office.

U.S. General Accounting Office. 1992. *Unemployed Parents.* GAO/PEMD-92-19BR. Gaithersburg, Md.: U.S. General Accounting Office.

U.S. House of Representatives. 1993. Committee on Ways and Means. *1993 Green Book.* Washington, D.C.: U.S. Government Printing Office.

Wallsten, T., and D. Budescu. 1983. "Encoding Subjective Probabilities: A Psychological and Psychometric Review." *Management Science,* 29: 151–173.

Webster's Eighth New Collegiate Dictionary. 1985. Springfield, Ill.: Merriam-Webster.

Westoff, C., and N. Ryder. 1977. "The Predictive Validity of Reproductive Intentions." *Demography,* 14: 431–453.

Whitman, D. 1992. "The Next War on Poverty." *U.S. News and World Report,* October 5, p. 36.

Willis, R., and S. Rosen. 1979. "Education and Self-Selection." *Journal of Political Economy,* 87: S7–S36.

Woittiez, I., and A. Kapteyn. 1991. "Social Interactions and Habit Formation in a Labor Supply Model." Department of Economics, University of Leiden, The Netherlands.

Name Index

Subject Index